Problem Solving Models of Scientific Discovery Learning Processes

European University Studies
Publications Universitaires Européennes
Europäische Hochschulschriften

Series XI
Education

Série XI Reihe XI
Pédagogie
Pädagogik

Bd./Vol. 425

PETER LANG
Frankfurt am Main · Bern · New York · Paris

Peter Reimann

Problem Solving Models of Scientific Discovery Learning Processes

PETER LANG
Frankfurt am Main · Bern · New York · Paris

CIP-Titelaufnahme der Deutschen Bibliothek

Reimann, Peter:

Problem solving models of scientific discovery learning processes / Peter Reimann. - Frankfurt am Main ; Bern ; New York ; Paris : Lang, 1990
    (European university studies : Ser. 11, Education ; Vol. 425)
    Zugl.: Freiburg (Breisgau), Univ., Diss., 1989
    ISBN 3-631-42709-3

NE: Europäische Hochschulschriften / 11

D 25
ISSN 0531-7398
ISBN 3-631-42709-3

© Verlag Peter Lang GmbH, Frankfurt am Main 1990
All rights reserved.

All parts of this publication are protected by copyright. Any utilisation outside the strict limits of the copyright law, without the permission of the publisher, is forbidden and liable to prosecution. This applies in particular to reproductions, translations, microfilming, and storage and processing in electronic retrieval systems.

Table of Contents

1. Introduction 1
    1.1 The Discovery Problem: Learning About Optics in a
        Computer-Simulated Laboratory 1
    1.2 An Example For the Problem Solving View of Induction:
        The General Rule Inducer 4
    1.3 Focus on Task Analysis 6
    1.4 Analyzing Processes of Hypothesis-Driven Discovery 7
    1.5 Computer Models of Discovery Learning 9
    1.6 Theories, Models, Implementations 10

2. Theoretical Background 13
    2.1 Introduction 13
    2.2 The Problem Solving Approach To Human Cognition 13
        2.2.1 An information-processing model of concept formation 15
        2.2.2 Constraints on Induction 16
    2.3 Production Systems As Tools For Modeling Problem Solving
        and Learning 18
        2.3.1 Production Systems as Theories of Cognitive Architecture 20
        2.3.2 Rule Acquisition in Production Systems 21
    2.4 Hypothesis Testing and Evidence Evaluation: Experimental Research . . 23
        2.4.1 Hypothesis Testing 23
        2.4.2 Evidence Evaluation 24
    2.5 Hypothesis Generation and Testing as Problem Solving 25
        2.5.1 Automated Scientific Discovery 25
        2.5.2 Condition Finding 28
        2.5.3 Learning By Analogy 29
    2.6 Conclusions 30

3. The Discovery World Program REFRACT 31
    3.1 Introduction 31
    3.2 The Domain: Geometrical Optics 31
    3.3 Running Experiments and Stating Predictions 34
        3.3.1 Conducting Experiments in REFRACT 36
        3.3.2 Gathering and Analyzing Experimental Data 43
    3.4 Gathering and Recording Data from the Student 46
    3.5 Conclusions 47

4. A Study of Learning in REFRACT 48
    4.1 Introduction 48
    4.2 Subjects and Method 48
    4.3 Observations From An Exploratory Study 49
        4.3.1 Learning Effects and Final Knowledge States 49
        4.3.2 Prediction Categories 52
        4.3.3 Discrimination of Relevance 53
        4.3.4 Design Selection 58
        4.3.5 Determining Covariation 59
        4.3.6 Response to Guidance-Free Learning 61

  4.3.7 Summary 61
 4.4 Analysis of Verbal Hypotheses 62
  4.4.1 Overview of S03's and S10's Learning Behavior 62
  4.4.2 Search Through The Experiment Space 70
  4.4.3 Hypothesis Modification 72
  4.4.4 Summary 77
 4.5 Conclusions 79

5. Discovery Learning as Problem Solving - A Task Analysis 81
 5.1 Introduction 81
 5.2 Constraints Based on Empirical Observations 81
 5.3 Extending the General Rule Inducer 82
 5.4 Search in the Experiment Space 86
 5.5 Search in the Hypothesis Space 89
  5.5.1 Dimensions of Search 91
  5.5.2 Generating Phenomenon Descriptions 93
  5.5.3 Modifying Hypothesis Scope 99
  5.5.5 Summary 104
 5.6 Conclusions 105

6. Computer Models of Hypothesis-Driven Discovery Learning 108
 6.1 Introduction 108
 6.2 A Quantitative Discovery Model: HDD-SH 108
  6.2.1 Model Description 109
  6.2.2 Learning Results and Learning Process 120
 6.3 Constraining HDD-SH 125
  6.3.1 Restricting Function Induction
    By Preferring Operators: HDD-SHR 127
  6.3.2 Restricting Function Induction
    By Preferring Operators and Variables: HDD-SHLR 128
 6.4 Learning variable and operator preferences: HDD-SHOP 130
 6.5 Summary: Quantitative Discovery Models 138
 6.6 Generating Qualitative Hypotheses 141
  6.6.1 Preference For Pictorial Descriptions 141
  6.6.2 A Pictorial Description Language 142
  6.6.3 Effects of Representation on Inductive Problem Solving 143
  6.6.4 A Qualitative Version of HDD: HDD-QUAL 145
 6.7 Why Do Learners Prefer Pictorial Representations? 148
 6.8 Changes in Problem Representation and the Case of
   Semi-Quantitative Hypotheses 149
 6.9 Conclusions 150

7. General Discussion 152

References 156

Appendix IV.1 162

Appendix VI.1 166

Appendix VI.2 177

Appendix VI.3 184

Appendix VI.4 192

Abstract

In order to study the mechanisms that underlay peoples' ability to generalize from specific observations and to use these generalizations to derive predictions, I developed a computerized discovery learning environment for geometrical optics, REFRACT, where predictions (about refraction phenomena) can be improved by analyzing specific observations made in the context of simulated (ray) experiments. This induction problem calls for applying basic 'scientific' discovery skills: hypothesis generation, design of experiments, derivation of predictions, comparison between predicted and observed changes.

To understand how learners with limited knowledge about the optics domain develop initial hypotheses about the laws governing the domain and how these hypotheses are corrected and refined under the influence of feedback, I analyze the discovery learning task from three different but related perspectives: Empirically by analyzing data from an exploratory study; conceptually by describing the discovery behavior as information processing based on theoretical notions from problem solving and learning research; and computationally by developing and studying computer models of learning and discovery.

In the empirical study, I observe how eight undergraduate students with no formal training in the sciences make discoveries in REFRACT. The observations are analyzed in terms of characteristics of discovery learning strategies: skill in constructing experimental designs, discriminating relevance of variables in REFRACT, determining covariation, and in terms of characteristics of the knowledge acquired: Final hypotheses hold by the subjects, degree of precision of their predictions, and so on. The study is exploratory in nature: It is meant to help me formulate specific hypotheses about discovery learning, not to test any.

With the conceptual analysis of the discovery task, I attempt to bring together the specific observations I made in the context of the empirical study with general notions of inductive reasoning. This analysis takes the form of developing an abstract information-processing model for the task that is based on the view of induction as problem solving and that obeys at the same time constraints derived from the empirical study. A knowledge representation language is employed that is particularly well suited to capture hypotheses about changes in an environment: rules. A predictive hypothesis about the path of rays in REFRACT is formally notated as a condition-action rule, where the condition part describes the scope of a hypothesis, i.e., the kind of experiments it can be applied to, and where the action part describes a refraction phenomenon. Based on this knowledge representation language, learning by discovery in REFRACT is described as rule induction. Rule induction comprises two problems: induction of phenomenon descriptions and restriction of a rules condition part. The issue of finding conditions for hypothesis-rules is important since in REFRACT hypotheses need to be generated before all the experiments are known to the learner and will therefore often be too general. The learning mechanism I suggest to correct for overly general rules is *discrimination learning*.

Finally, I develop a computational model of the discovery task in form of a computer program written in a production system language, that is, realized as a rule-based program. Assumptions about the cognitive architecture of a discovery learner are represented in form of the production system architecture, assumptions about problem representation are expressed in terms of a specific description language for experiments, the knowledge required to perform in REFRACT is stated in form of production rules that manipulate descriptions of the environment and internally considered goals, and a discrimination learning algorithm is employed to model condition modification. I further experiment with specific variants of the computational model, where each variant stands for a specific approach to the discovery problem. The models can be classified along two dimensions: Whether they use a more quantitative or more qualitative representation of experiments and hypotheses, and, within the quantitative versions: Which kinds of constraints are imposed on the function generation step. These dimensions are derived from my observations on human discovery learners. I compare the model variants with respect to their discovery behavior and with respect to the resulting knowledge structure.

I claim that these kinds of 'computational experiments' are an effective way to generate detailed hypotheses concerning the complex interactions between cognitive architecture (e.g., memory limitations), several kinds of 'knowledge' a learner might have, specific learning mechanisms, and the dynamically changing learning environment. Further, that this approach can be used to evaluate the plausibility of such hypotheses. Finally, I discuss the usefulness of this kind of analysis of a discovery learning task in the context of instructional applications, in particular for building intelligent tutoring systems.

Acknowledgements

This work would not have been possible without the support from some very special people. First to mention is my thesis advisor, Hans Spada, who initiated my interest in cognitive psychology and has helped me over the years more than once to see the forest again when I only saw trees. And without the steadily improving computational facilities in his laboratory, many parts of my project could not have been accomplished. Another person that had significant influence on this thesis is Robert Glaser, with whom I had the luck to work for two years. His personal advice and the great support I received at the Learning Research and Development Center, which he co-directs, helped me to realize things that otherwise would not have been possible.

Also at LRDC in Pittsburgh, it was Kalyani Raghavan and Jamie Schultz who provided me with valuable conceptual and technical suggestions on how to construct and improve the REFRACT program. In Freiburg, the theoretical and technical expertise of my colleagues Klaus Opwis, Rolf Plötzner, and Michael Stumpf, not to mention their personal support, helped me in many ways and stimulated my work on the computer modeling issues.

Last, but certainly not least, I want to thank Fieny, who has not only been a keen critic of my scientific prose, but also a very close friend when I needed it most.

List of Figures

Fig. 1.1: Generators in a concept formation task 5
Fig. 2.1: Main components of a production system interpreter 19
Fig. 2.2: Control structure of a production system interpreter 20
Fig. 3.1: Points, distances and angles in REFRACT 32
Fig. 3.2: Different surface forms 34
Fig. 3.3: REFRACT's main screen 36
Fig. 3.4: The menu used to select a medium type 37
Fig. 3.5: Menu to select the object distance 38
Fig. 3.6: Menu to select the angle Alpha 38
Fig. 3.7: A completed design 38
Fig. 3.8a: A RayArea prediction 40
Fig. 3.8b: A ImageArea prediction 40
Fig. 3.8c: A Ray prediction 41
Fig. 3.8d: Distance and angle prediction 41
Fig. 3.9: The screen after feedback is displayed 42
Fig. 3.10: The LabWindow with attached inspection tools 43
Fig. 3.11: The NoteBook 44
Fig. 3.12: A subtable with variables 45
Fig. 3.13: The Psychologist Window 46
Fig. 4.1: Prediction type (Area, Ray, Number) selection 54
Fig. 4.2: Graphical predictions made by Subject S03 65
Fig. 4.3: Design Sequence of Subject S03 70
Fig. 4.4: Design Sequence of Subject S10 71
Fig. 4.5: Verbal design reasons 72
Fig. 4.6: Prediction type preferences 73
Fig. 4.7: Frequency of looking for numerical relations 74
Fig. 4.8: Connections between variables for S03 and S10 78
Fig. 5.1: The main conceptual components of HDD 86
Fig. 5.2: Prediction, hypotheses, and production rules 90
Fig. 5.3: Outline of the function induction step 96
Fig. 5.4: Essentials of the discrimination step 101
Fig. 6.1: The goal sequence realized by HDD-SH 110
Fig. 6.2: Production system architecture of HDD-SH 112
Fig. 6.3: The interaction between memory WM and ENV 113
Fig. 6.4: Hypothesis selection during the prediction step 114
Fig. 6.5: Flow of control in HDD-SH 121
Fig. 6.6: Discrimination tree for HDD-SH 122
Fig. 6.7: Development of number of hypothesis-rules 124
Fig. 6.8: Development of number of elements in working memory 124
Fig. 6.9: The 'Bucket Brigade' mechanism 131
Fig. 6.10: Architecture of HDD-SHOP 132
Fig. 6.11: Flow of control in HDD-SHOP 134
Fig. 6.12: Interaction of memories CONTROL and PHEN 136
Fig. 6.13: Trace of HDD-SHOP 139
Fig. 6.14: Three different reference lines 143
Fig. 6.15: Example for ray paths 147

1. Introduction

How do people discover the concepts, regularities and laws that govern a certain aspect of their environment with little or no explicit instruction? Such regularities are often hard to discover because observations are not gathered systematically or because they contain noise, because factors influencing the relations are not known and hence not controlled, or because the discovery learner does not know how to parsimoniously describe a body of observations. In order to deal with a potentially overwhelming amount of observations, people develop *hypotheses* about the laws that control a domain and test these hypotheses by deriving predictions and comparing predicted changes with observed changes in the environment. Hypotheses are thus employed to structure experience and to direct inferencing, or, as John Locke already put it:

> Hypotheses, if they are well made, are at least great helps to the memory and often direct us to new discoveries.

The topic of this thesis is the *acquisition and refinement of hypotheses* in the context of a scientific discovery learning task. The learning is called 'scientific' because a scientific concept, optical density, and a scientific law, Snell's Law (also known as the Law of Refraction) are to be discovered. I speak of scientific discovery 'learning' instead of the more general 'scientific discovery' because I do not analyze the professional scientist's discovery skills, but those of scientifically untrained adults.

1.1 The Discovery Problem: Learning About Optics in a Computer-Simulated Laboratory

I became interested in processes of discovery learning and reasoning by hypotheses as a student at the University of Freiburg, when I was involved in a developmental study, inquiring how children of various age groups acquire knowledge about optics in a self-guided manner (Spada, Reimann & Häusler, 1983). While we had developed interesting strategies to present the domain to subjects, they were technically difficult to accomplish. Further, in order to record the dense trace of subjects' discovery behavior required for our analysis, two trained experimenters were needed to work with a single subject. The idea early arose to present the learning task on a computer and have the machine take care of recording subjects' behavior.

The opportunity to develop such a computerized discovery environment for optics came when I joined a group at the Learning Research and Development Center (LRDC), University of Pittsburgh, that had the goal to develop computerized instructional microworlds. I developed the computer program REFRACT (Reimann, 1988a) with the intention to have a tool that allows to study some central components of (scientific) discovery learning in a somewhat natural context, but being at the same time able to trace students' discovery behavior in a detailed manner: hypothesis generation, design of experiments, testing hypotheses by deriving predictions and comparing them with feedback information. In REFRACT, students can learn about a simplified version of Snell's law (law of refraction) by running experiments, predicting their outcome, and processing information about the correct outcome of experiments as displayed by the program. Note that the environment provides the discoverer with 'feedback' that contains more than just categorical information such as *right* or *wrong*. No additional instruction is provided about the domain. The student working with REFRACT has to induce regularities by generalizing from the experiments he[1] runs. Snell's Law describes the phenomenon that light rays are refracted ('bent') when they go from one optical medium (such as air) into another medium with a different optical density (for example, water). The law describes the amount of refraction as the relation between the *angle of refraction*, the *angle of incidence*, and the difference in the *optical density* of the two media. The domain and the way it is presented to students is described in Chapter 3.

Although REFRACT constitutes a specific discovery problem, it allows me to address three aspects of scientific discovery learning that are of general importance: data collection by means of experimentation ("I designed this experiment in order to find out whether substance type has any effects on the degree of refraction"), generation of hypotheses that take the form of descriptive generalizations ("The object distance will always be two times the image distance"), and testing of hypotheses by means of deriving predictions and comparing them with experimental results ("The object distance will be 200 units in this experiment"). The focus is on *descriptive hypotheses*, that is, parsimonious descriptions of a set of observations. It is not on explanatory (or causal) hypotheses such as "The ray is bent when it enters the glass body because the glass exerts some force on it".

Students that work in the discovery environment REFRACT are confronted with the following problem: They have to state predictions about the path of refracted rays based on *insufficient knowledge*. It is insufficient in two ways: Students do not know about the law of refraction, so their predictions can only be based on the observations they conduct in the learning environment. Insufficient also because students in REFRACT cannot run experiments and just observe what happens, thus waiting with formulating predictions until they have gathered enough information to formulate somewhat save generalizations. Instead, they have to deliver a prediction for every experiment they run. It is my basic assumption that students react to this demand by formulating hypotheses (i.e. assertions which are not necessarily true) about the regularities that govern the behavior of rays in the particular experiments, and that they test these hypotheses by designing new experiments, predicting the ray path based on a currently believed hypothesis, and comparing the predicted ray path with the correct one calculated by the program. This comparison will have repercussions on the knowledge state: A hypothesis may be modified

---

[1] Masculine expressions are used as generic terms; no bias is intended.

because it led to a wrong prediction, new hypotheses may be generated, or people may become more confident about a hypothesis that resulted in a correct prediction.

Even though experiment generation is part of the discovery task introduced in this thesis, I will for the most part concentrate on hypothesis generation and testing. I do not make a strong distinction between generation and testing of hypotheses. Rather, hypothesis testing is seen as a process comprising the derivation of a prediction about changes in the environment, the comparison of predicted with observed changes, and, finally, the modification of the hypothesis. Thus, hypothesis testing is closely tied to hypothesis generation in that it leads to the modification of an incorrect hypothesis. Questions of how people with limited knowledge about a domain develop initial hypotheses about the laws governing the domain, and how hypotheses are corrected and refined under the influence of informational feedback have rarely been analyzed in psychology, despite their central importance for processes of discovery and induction in general. Also, little is known about the nature of differences between humans with respect to these skills.

The strategy to describe discovery learning as hypothesis testing is supported both by theoretical positions in cognitive psychology and by empirical observations. As for the theoretical groundwork, consider the following statement:

> Representational theories of mind presuppose hypothesis-testing theories of learning. In explaining the acquisition of any body of knowledge, one must specify the class of hypotheses that the organism entertains, the evidence that is taken into account as relevant to decisions among hypotheses, and evaluation metrics for that evidence (Carey, 1984, p. 47).

Thus, the notion of mentally entertained hypotheses is one of the core constructs in cognitive psychology. The hypothesis metaphor further permeates virtually all work on concept formation (e.g., Bruner et al., 1956,; Bower & Trabasso, 1964; Levine, 1966; Medin & Smith, 1984), as it is documented in Chapter 2 where I review some of the relevant research literature. Besides giving a short overview of experimental research on induction and scientific reasoning, I concentrate on theories and methods developed in the more computational branches of cognitive psychology. Since I analyze the particular discovery learning task within a specific theoretical and methodological framework: *problem solving theory* and *cognitive modeling* using *production systems*, discussion of the theoretical background is for the most part restricted to these approaches and related areas from Artificial Intelligence research. Inductive and scientific reasoning is seen as a special kind of problem solving. This view was originally put forward by Simon & Lea (1974) who suggested that both problem solving tasks and concept formation tasks can be carried out by a single information processing system. Simon and Lea's information-processing analysis of inductive reasoning is a good example for the kind of analysis I want to provide for a specific discovery learning problem.

1.2 An Example For the Problem Solving View of Induction: The General Rule Inducer

The basic idea of Simon and Lea's (1974) proposal is that concept formation (Bruner, Goodnow & Austin, 1956; Levine, 1966) and inductive thinking in general can be seen as just another kind of search process and are thus closely related to problem solving. The two task classes are similar as far as the same general methods are used to solve concept formation (or rule induction) tasks as well as standard problem solving tasks. These methods are outlined by the authors in form of a conceptual computer model called the General Rule Inducer (GRI). Problem solving and rule induction are different from each other with respect to the kind of problem spaces that are searched. Rule induction requires to search in two spaces, a space of *stimuli*, of *data*, or of *instances*, and a space of possible *structures*, such as *rules*, *patterns*, or *relations*: "In a problem of induction, some material is presented and the problem solver tries to find a general principle or structure that is consistent with the material" (Greeno & Simon, 1984, p. 82). Ordinary problem solving can be accomplished within a single problem space.

To understand the duality of problem spaces, we have to recapitulate the general nature of any problem solving process, including rule induction (cf. Newell & Simon, 1972). Problem solving - seen as information processing - requires two basic modules: There must be a *generator* to produce problem states, and a *test* module telling whether the goal state is included in the problem states generated. In the context of 'normal' problem solving, the generator and the test module apply to states contained in a single problem space, since the goal state description is part of the problem space. For example, in the Tower-of-Hanoi problem the goal state might be encoded as *(Disc-1 and Disc-2 and Disc-3 on Peg-C)*, and the current problem solving state might be described as *((Disc-1 and Disc-2 on Peg-C) (Disc-3 on Peg-B))*. Comparing these two symbolic descriptions might result in two things: For one, the executive realizes that the two states are not identical. Secondly, if the problem solving system is equipped with knowledge about how to employ the so-called means-ends-strategy, it might realize that moving *Disc-3* to *Peg-C* is the next promising goal.

The situation is different for rule induction. Here, no description of the goal state exists. Rather, the goal state is formulated as a constraint on the rules: It should be a rule that classifies all instances correctly. Rules take the form of, for example, classifications: *IF (Instance Round) And (Instance Green) THEN (Instance is member of the class)*. A generator for inductive hypotheses produces rules of this kind. In order to test whether such a rule is the goal state, the rule must be matched against instances. However, descriptions of instances are not part of the rule space: they cannot be generated by applying rule modification operators to hypotheses. Rather, they form a different, second problem space, one with its own (instance) generator[2]. Testing an inductive rule

---

[2] The experimenter might take the role of the instance generator, for example, in the reception paradigm of concept formation experiments.

requires matching it against the elements from the instance space, hence, against another problem space. In terms of search strategies employed, rule induction is based on a generate-and-test cycle with "separate generators for rules and instances" (Simon & Lea, 1974, p. 115). In order for induction to work, the two spaces have to be connected.

Based on this distinction between a rule space and an instance space, a variety of problem solving strategies for rule induction problems can be conceptualized. They are distinguished according to the relation between the rule generator, the instance generator, and the nature of the test process. Figure 1.1 depicts the components and their interrelations. In the most primitive case--*generate and test*--there is no feedback from test to the rule generator (Channel *e*); the test eliminates rules, but does not provide the generator with information on how to construct more promising rules. A more clever human or artificial system could use information from the test process to modify existing rules (Channel *a* and *e*). Simon and Lea label this strategy 'the rule induction version of the heuristic search method'. Further problem solving strategies for rule induction can be employed. The rule generator might be interfaced with the instance generator so that it can drive instance creation (Channel *a*): *heuristic search for instances*. Or information from the instance space (not only from the recent test) can be used to drive the rule generator (Channel *b*): *heuristic search for rules*.

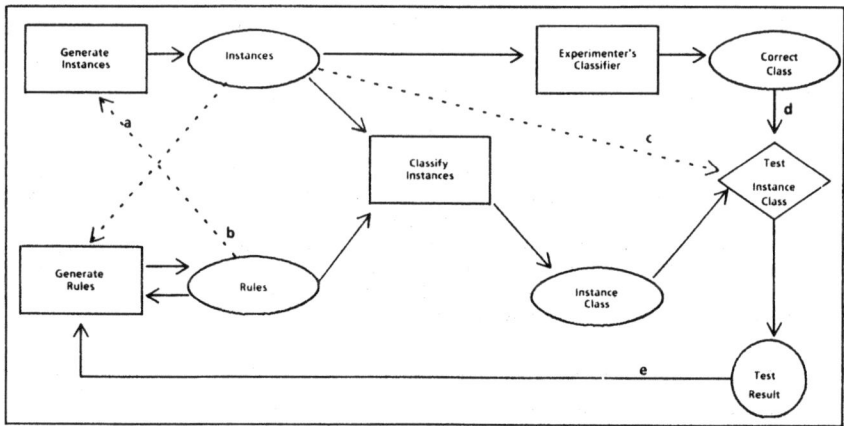

Fig.1.1: Generators in a concept formation task (after Simon & Lea, 1974)

The search methods outlined in GRI are not confined to the realm of concept formation tasks, but work equally well for problems of sequence extrapolation (Simon & Kotovsky, 1963; Kotovsky

5

& Simon, 1973), grammar induction (Siklossy, 1968), induction of mathematical functions (Langley, 1979b) and aspects of scientific discovery (Langley, 1979a; Langley et al., 1987; Klahr & Dunbar, 1988). In later chapters, I will extend the framework provided by Simon and Lea's analysis so that the discovery task I shall investigate can be described as a search in two problem spaces, one for experiments and one for hypotheses. The problem solving view as illustrated with GRI serves as a framework with respect to the theoretical concepts I employ - discovery learning is treated as a rule-induction task and, ultimately, as problem solving -, and on behalf of the methods I will utilize: analysis of verbal data, conceptualization of information processing models for the task ('task analysis'), and development of implemented computational models of learning and performance. Modifications and at times extensions of the conceptual GRI model will be required to adapt the rule induction framework to the discovery task at hand; and I will move from the level of conceptual considerations about information-processing capacities to running computer implementations.

1.3 Focus on Task Analysis

My goal is to analyze knowledge acquisition by discovery learning where the type of analysis I'm striving at is best characterized as task-analytical. With task analysis I mean a detailed description of the mechanisms that are *sufficient* to solve the discovery problem in REFRACT. Detailed in the sense of being precise enough to be realizable as a computer program. These mechanisms should be *psychologically plausible* inasmuch as they should comply with empirical findings and with theories about human inductive thinking, learning and memory. That is, the computer program based on such a task analysis should produce behavior that reveals a general similarity to performance of human problem solvers in the discovery task. The general motivation for task analytical concerns is expressed concisely by Simon (1975):

> Discovering of what subjects learn can be approached experimentally, but important preliminary insights can be gained by analyzing the structure of the task itself to determine the possible alternative ways of performing it. (...) Different subjects may in fact learn different things in the same task environment, and a formal analysis of the environment can help define the range of possibilities. It can also help define the differences in demands that different methods of task performance place upon the subject" (p. 230).

With characterizing my approach as task-analytical I want to draw two demarcation lines. The first one in order to distinguish the procedure from experimental research: I did not sample subjects and put them into an experimental setting, designed to vary some factors and to keep others constant. My research is empirical in as far as I develop models of discovery behavior that are partially based on observations on human students and in as far as these models can be used to derive predictions about the discovery behavior of human subjects. However, these predictions do refer to specific statements of single discovery learners, not to statistics that summarize data aggregated over a group of subjects. The second demarcation line is necessary to distinguish the computer models of discovery learning I developed from *simulation* models. In cognitive psychology, with a simulation model one usually refers to a computer model that simulates the performance of a single subject with respect to some cognitive task. My computer models are not simulation models in this sense, since they do not mimic the behavior of a single subject. Rather, they are models of idealized, prototypical discovery learners.

1.4 Analyzing Processes of Hypothesis-Driven Discovery

Both general accounts of inductive thinking (e.g., Simon & Lea, 1974; Holland, Holyoak, Nisbett & Thagard, 1986) as well as specific analysis of scientific discovery tasks (e.g., Mynatt, Doherty & Tweney, 1978; Langley, 1979a) are either to global or to context-specific as to allow for a direct application to the discovery task in REFRACT. I therefore undertook an empirical study with the intention to develop more specific hypotheses about how learning in REFRACT does take place (Reimann, 1988b). The main results are described in Chapter 4. This study was exploratory in nature, meant to help me generate hypotheses, not to test any. I analyze subjects discovery behavior descriptively and provide for two subjects a more in-depth analysis, taking into account their verbal utterances.

The observations I made on students working in REFRACT influenced the conceptual and computational models described in later chapters in form of *constraints* on those models. The central constraints derived from the empirical study are: (a) do not assume that experiments are designed systematically; (b) do not assume a lot of knowledge about data analysis methods; (c) don't expect that hypotheses are consistent with all the evidence and with respect to each other; (d) do not assume that significant causal knowledge about optics is brought to the task; and (e) do not assume that experiments are necessarily represented in a numeric format.

Chapter 5 is devoted to a task analysis that has as its goal the development of a *conceptual* model of the discovery task. The conceptual model of the task is called HDD[3], for *Hypothesis-Driven Discoverer*. In order to make the notion of 'hypothesis' and 'knowledge structure' more concrete, I use a knowledge representation language that is particularly well suited to capture conditional assertions about changes in an environment: rules. A predictive hypothesis about the path of rays in REFRACT is formally notated as a condition-action rule, where the condition part describes the *scope* of the hypothesis, i.e., the kind of experiments it can be applied to, and where the action part describes a *refraction phenomenon: If the medium has a plane surface, then the ray goes straight through.* I see discovery learning as acquiring and modifying rules. Rule sets are the central knowledge structure in my analysis. Means to generate new rules, to modify incorrect rules, and to integrate new and modified rules into the existing knowledge structure comprise the

---

[3] In this text, I will use the acronym HDD to refer to the conceptual, i.e., not implemented, information-processing model of discovery learning. Actual implementations will have a label that carries a specific index, for example, HDD-SH, HDD-SHR.

kind of learning focused on.

Based on this knowledge representation language, learning by discovery in REFRACT is described as rule induction in the sense Simon & Lea (1974) use this term. Rules are induced as a result of search in two problem spaces, one for experiments and one for hypotheses. A formal characterization of the two search spaces, the experiment space and the hypothesis space, is provided. HDD comprises the following specifications: Experimental data, the elements of the experiment space, are notated as feature vectors; hypotheses, the elements in the hypothesis space, are expressed as rules. It includes further specifications for three kinds of cognitive operators: For the kind of knowledge required to design an experiment, for constructing a hypothesis, and for modifying a hypothesis. Issues of experiment generation are not treated in much depth. Rather, my focus is on the search through the hypothesis space. The problem of searching for good hypotheses (i.e., those that allow precise and correct predictions) is divided into two subproblems: search for condition ('scope') descriptions and search for action ('phenomenon') descriptions.

Phenomena descriptions for REFRACT capture regularities about the path of refracted rays. Subjects in my empirical study stated regularities in assertions like: *The ray is bent upwards*, *As the object distance increases, the image distance decreases*, or: *The angle of refraction equals two times the angle of incidence*. Thus, they phrase refraction phenomena in 'pictorial' form describing the path the refracted ray will take, or in semi-quantitative form, or in form of functional relations between dependent and independent variables. The conceptual model of learning in REFRACT covers only phenomena descriptions of the third, quantitative type and says indirectly something about how semi-quantitative hypotheses are developed. My initial analysis will have little to say on qualitative hypotheses. I return to this issue in Chapter 6, where I make some suggestions of how to go with the problem and where I provide a first implementation of a 'pictorial' discovery learning system.

The issue of finding conditions for hypothesis-rules is so important in HDD since hypotheses need to be generated before all the experiments are known to the learner and will therefore often be too general. The learning mechanism I employ to correct for overly general rules is *discrimination learning*. That means, the scope of a hypothesis that led to a wrong prediction will be constrained (specialized) by adding conditions to the hypothesis. Generation of rules on the basis of incomplete knowledge and revision of hypotheses that are no longer in accordance with the data are the central concerns of my analysis. For the phenomenon induction problem as well as the condition modification task the guiding constraint is that learners trade ease of computation for completeness and consistency, a strategy I found in my subjects, which is also in accordance with general theories of inductive reasoning (Holland et al., 1986; Langley, 1987).

1.5 Computer Models of Discovery Learning

Based on the conceptual information-processing analysis of the task I implemented *computer models* of discovery learning in REFRACT. Chapter 6 is devoted to a description of these implementations. They take the form of *production system* programs written for the production system shell PRISM2 with considerable support in form of Lisp functions. The step from the conceptual information-processing model of discovery learning to the computer implementations was required because the processes that are needed to generate model performance interact in such a complex manner that it is no longer possible to predict what the - only verbally described - conceptual model will discover, given a certain constellation of its internal parameters and a specific experiment sequence.

Except for design construction[4], the computer models accomplish the main components of discovery in REFRACT: They induce phenomena descriptions based on specific experiments and capture them in hypotheses, they derive a prediction by applying a hypothesis to a design description, they evaluate the hypothesis by comparing prediction with feedback, and they correct for overly general hypothesis by attaching new conditions to them. I thus show how processes of hypothesis generation and modification in REFRACT can be realized computationally. This is done by expressing assumptions about the cognitive architecture of a discovery learner in form of a production system architecture, representing assumptions about problem representation in terms of a specific description language for experiments, representing knowledge about how to perform in REFRACT in form of production rules that manipulate descriptions of the environment and internally considered goals, and by implementing learning mechanisms in form of production rules and Lisp code. All models are built around an architectural core that comprises four production system memories. There is a kind of external memory representing experiments, distinguished from a second, more limited, working memory; and a distinction is made between knowledge *used* for discovery learning and knowledge *acquired* through discovery learning: separate production memories are defined for the two components.

In the first computer implementation provided, phenomena descriptions take the form of equations relating a dependent to an independent variable. I start with a 'numeric' discovery learner since this type is much easier to handle than 'qualitative' discovers. The first implementation corresponds pretty much to the model of a discovery learner as it emerged from the task analysis in Chapter 5. This discovery learner is besides other things characterized as inducing functional relations between variables in REFRACT in an unsophisticated way relying on a generate-and-test strategy. This results, not surprisingly, in an 'explosion' of the number of hypotheses generated, but the correct hypotheses are found.

The next two model versions I introduce also work in a 'numeric' mode. They model learners

---

[4] Experiments are provided as input to the programs.

that use more specific heuristics to induce functions. Application of these heuristics does reduce the number of implausible hypotheses generated by the system. However, these heuristics are fairly ad-hoc or task-specific, respectively. I continue by describing another version of a discovery leaner that is more general in that it does not use specific function-induction heuristics right from the start (as if they would be 'wired in'), but *learns* to prefer certain function-induction operators and variables based on its success in predicting ray paths. Thus, it learns not only about which hypotheses to prefer given success and failure in prediction making, but it learns also to select promising variables from REFRACT and to select promising mathematical operators to construct predictive equations. This second kind of learning ability is accomplished by a mechanism for distributing credit and blame not only over hypothesis-rules, but also over the rules that are involved in constructing hypothesis-rules.

I then turn to the question of how to deal with the problem of the qualitative, more precisely, 'pictorial' hypotheses within the computational framework developed so far. After some more theoretical considerations about the possible effects a pictorial representation of objects and relations in REFRACT can have on the discovery process and its outcomes, a more suggestive than decisive first implementation of a 'pictorial' discovery learner is described. With this I go beyond the task analysis provided in Chapter 5.

1.6 Theories, Models, Implementations

I use the terms 'theory', 'model' and 'implementation' (or 'computer model') throughout the thesis without defining them precisely; I'm afraid I do not even distinguish between them consistently. In general, this is due to the fact I do not want to draw a clear line between these three terms, primarily because the issue of whether such demarcation lines should be drawn and if so, along which coordinates, are topics of an open debate in cognitive science (for some contributions to this debate see Kintsch, Miller & Polson, 1984; Pylyshyn, 1984). However, a word on some tentative distinctions between the three notions is in order.

I use the term *theory* in this thesis almost only in the context of a specific theory: *problem solving theory* as developed in Newell & Simon (1972) and Anderson (1976, 1983), to mention the main protagonists. I do not make any significant distinction between problem solving theory and information-processing theory of human cognition, since both can be traced back to the same sources. Problem solving theory, as summarized in Chapter 2, is used as the background theory of this thesis.

When speaking of *conceptual models* I refer to applications of problem-solving (or information processing) theory to a specific problem or class of problems. For example, the General Rule Inducer introduced before is a conceptual model that resulted from applying general concepts of problem solving theory to the class of rule-induction problems. The conceptual model that is developed in Chapter 5, HDD, is an application of problem-solving constructs to a specific discovery problem, the one present in REFRACT. Such models are qualified as 'conceptual' since they describe task characteristics in terms of problem-solving and, hence, information processing, but not in form of a computer program. What does a conceptual model 'model'? Usually, it is a kind of 'ideal' problem solver that is represented.

Implemented models of a task are computer programs that are executable and can solve the problem or class of problems they refer to. An implemented model can represent an abstract, ideal information-processing system, or it can model a specific problem solver (with respect to a particular task or task class). When a computational models refers to a specific human being it is usually called a *simulation model*. The model described in Newell (1967) is a famous example for a simulation model.

Connected with the steps from theory to conceptual model to implementation is usually a loss in generality and a gain in precision and testability. Existing computer implementations of cognitive capacities can hardly be generalized in the sense that one and the same program could solve problem from different domains. However, adapting a general theory to a specific task and making it so concrete that it can be programmed and then executed on a computer helps a lot in order to decide whether the theory provides one with the sufficient means to explain behavior in a specific task.

Using computer simulation programs does not automatically guarantee that the complexity problem will disappear. What happens often is that one black box (the human cognitive machinery) is replaced with another black box, a computer program. Having to accept an intricate program as a model of some aspect of human intelligence makes it hard for the reader to grasp what the general theoretical principles behind the model are, and what does have to count as mere technical support. Furthermore, given a specific cognitive simulation program it may be difficult or impossible to decide what is true for all subjects and what is meant to be true for a particular subject. One has to identify those parts of a simulation program that constitute invariant features and has to specify how the (non-numeric) parameters of the models can be tuned to account for interindividual differences. The question of tuning parameters is the so called *tailorability* problem (VanLehn, Brown & Greeno, 1984).

I attempt to discriminate between the more principled components in my computational models and the many components that are introduced for mere technical reasons by distinguishing between the conceptual model of the task and the implemented models of the task. The conceptual model is supposed to describe the principles (such as heuristically controlled phenomenon induction, discrimination learning, competition between hypothesis-rules) without having to refer to constraints that are only technically motivated. Of course, when developing the conceptual task model I had a specific implementation language in mind: production system programming. However, as will be explained in Chapter 2, the use of production systems can to a large degree be justified based on general arguments about the nature of human information processing and learning (such as its pattern-matching nature, memory limitations, control characteristics, incremental nature of learning) and does therefore not count as a mere technical decision.

In order to enable the reader to estimate which parameters of my models need to be tuned in order to account for specific differences between human discovery learners, I decided not to present a single implementation that miraculously solves the problem in a way typical for certain subjects. Instead, I present in Chapter 6 a series of program versions, where each version is described in form of modules comprising 'architectural' assumptions and 'knowledge' assumptions. This way, it can easily be seen which changes were necessary to accomplish a certain characteristic of the modeled discovery behavior.

2. Theoretical Background

2.1 Introduction

I provide a selective review of research considered most relevant for my main goal: To analyze hypothesis-driven discovery learning processes in terms of search. I begin with summarizing the essentials of the approach that introduced the notion of search into psychology: problem solving theory, and by relating this approach to questions of inductive reasoning. Next, the specific programming formalism I will use in later chapters is considered: production systems, and the close relation between this formalism and problem solving theory is explained. I then inspect some of the classical research on induction and scientific reasoning and explain why most of this research is not directly applicable to more complex discovery tasks. Finally, computer programs are introduced that do some kind of automated scientific discovery and can learn from examples, building on the problem solving/search view of induction.

2.2 The Problem Solving Approach To Human Cognition

From current accounts of problem solving in cognitive psychology (Newell & Simon, 1972; Anderson, 1983; Greeno & Simon, 1984; VanLehn, 1988) the following essentials can be abstracted:

- The human brain can be understood as an information processing system. Thinking is accomplished by manipulating symbols in memory: receiving, comparing, copying, reorganizing, outputting symbols.

- Problems are solved by creating and manipulating a *problem space*, a symbolic representation of the problem comprising initial, intermediate and final problem states as well as all the other information required for problem solving. The problem space is changed dynamically by applying *operators* to the symbolic description of states contained in it. This corresponds to a mental *search* through the problem space.

- The search through the problem space for a problem solution is guided by *heuristics*, rules of thumb that control the application of operators, capitalizing on information about the structure of the problem space.

In other words, the cognitive theory of human problem solving has developed two kinds of concepts: hypotheses about the form of cognitive *action*, and hypotheses about the form of cognitive *representation* (Greeno & Simon, 1984). Hypotheses about cognitive action comprise domain-related action knowledge and strategic knowledge. For example, in proofing logic theorems, action knowledge takes the form of rules of inference, and strategic knowledge comprises heuristics such as working upwards from the givens, working downwards from the theorem to be proven, etc. The hypotheses about cognitive representation state that the problem solver has the problem situation represented in form of a problem space. With respect to logic problems, this means that he has a representation in memory encompassing the given axioms, the goal state to be reached (a theorem), and the intermediate steps of the current proof. Applying an operator to a problem state in this case means to apply an inference rule to a logic formula. For situations where more than one inference rule can be applied - and those are frequent - heuristic knowledge is necessary to decide which operator to apply.

Problem spaces are stored in memory. Two kinds of memories are distinguished: There is a data structure containing a practically infinite amount of information over very long times (long-term memory), and a second memory with very small capacity that holds information only for short time intervals (short-term memory). Symbolic manipulation of information can only take place in short-term memory, which is also called 'working memory'. Short-term memory serves as the interface between the sensory organs and long-term memory.

We said that heuristics utilize information from the problem space to guide the application of problem solving operators. Since a problem space may carry more or less information (have more or less structure), heuristics can be ordered along a dimension of reliance on information from the problem space (and, ultimately, from the task domain). On the one end of this dimension we find *weak methods* that do not depend on the specifics of a task and thus are very general. *Strong methods*, on the other hand, are much more effective than weak methods, but often bound to a particular problem or class of problems. An example for a strong method for discovering logic proofs is the strategy of 'proof by contradiction' that is specific to problems that have an axiomatic structure. The most important weak problem solving methods are:

- *Generate and Test* as the most basic method. It requires an operator that generates moves in the problem space and a second operator that tests whether a newly generated state constitutes a goal state. Variations of this method are based on the 'intelligence' built into the generator;
- *Hill-climbing* is a specialization of the generate and test method. Information from the test step is used to guide the generator in producing the next state;
- *Means-ends analysis* builds on heuristics to relate differences between current state and goal state to operators that can reduce the differences most effectively. This is very useful for problem classes where a clear definition of the goal state exists, for example, in discovering proofs in logic.

As I mentioned when introducing the General Rule Inducer in Chapter 1, the first two kinds of search control can be applied for rule induction problems, i.e., for induction, as well. The models I developed in the context of analyzing discovery learning do for the most part rely on a simple generate-and-test method, but two of them also utilize a kind of hill-climbing strategy (see Chapter 6). Let me summarize this exposition of essential problem solving concepts with a quote from Simon & Lea (1974):

> (1) There is a *problem space* whose elements are knowledge states. (2) There are one or more *generative processes* (operators) that take a knowledge state as input and produce a new knowledge state as output. (3) There are one or more *test processes* for comparing a knowledge state with the specification of the problem state and for comparing pairs of knowledge states and producing differences between them. (4) There are processes for *selecting* which of these generators and tests to employ, on the basis of the information contained in the knowledge states (p. 109-110).

Problem-solving models of induction have to be described as information processing systems, as systems that manipulate internal, symbolic descriptions of the environment and of the system's goals. To illustrate what it means to provide an information-processing account of inductive thinking, let me describe an attempt by Gregg and Simon (1967) to reformulate a stochastic theory of simple concept formation as an information-processing model.

2.2.1 An information-processing model of concept formation

Starting point of Gregg and Simon's considerations is Bower and Trabasso's (1964) mathematical theory of simple concept formation[1], which can be summarized as follows:

1. On each trial the subject is in one of two states, K or K'. If he is in state K (he "knows" the correct concept), he will always make the correct response. If he is in state K' (he "does not know" the concept), he will make an incorrect response with probability p.
2. After each correct response, the subject remains in his previous state. After an error, he shifts from state K' to K with probability q.

Bower and Trabasso mathematically derive from this theory expected values and variances for a large number of statistics which are confirmed by empirical data from experiments. In a sense, this theory can be regarded as a process model for subjects' behavior. The first statement describes the (probabilistic) response process, the second the (also probabilistic) learning process. It falls short as an information-processing model since it says nothing on the processes that *generate* the probabilities. However, as Gregg and Simon claim, the theory captured in statements (1) and (2) is based on assumptions about information processing, which are explicitly stated by Bower and Trabasso in formulations like:

> The subject in a concept-identification experiment is viewed as testing out various hypotheses (strategies) about the solution of the problem. Each problem defines for the subject a population of hypotheses. The subject samples one of these hypotheses at random and makes the response dictated by the hypothesis. If his response is correct, he continues to use that hypothesis for the next trial; if his response is incorrect, then he resamples (with replacement) from the pool of hypotheses.... (Bower & Trabasso, 1964, p. 39).

---

[1] 'Simple' means that the concept is defined by only one attribute, such as color = red.

Gregg and Simon continue by showing among other things that (a) without this or a similar background theory, formulation of the mathematical model is not possible; (b) while the stochastic theory can be derived from the background theory, the converse is not possible; (c) there is a many-to-one mapping from variations of the background theory to the stochastic models. This means: Psychological important modifications of the background theory make no difference in the stochastic model and can hence not be tested empirically by means of testing the stochastic theory. For example, it does not make a difference for the stochastic model that subjects instead of sampling from the complete set of hypotheses might sample form a set of hypotheses that does not include the last incorrect hypothesis. Obviously, a formalization of the background theory is desirable over its stochastic specialization. Gregg and Simon develop such a formalization in form of an information-processing model of the simple concept formation task, a computer program that does cover the background theory. They show that the Bower-Trabasso theory can be derived from that information processing model by means of aggregation, and demonstrate that their process model makes a better theory because it is more plausible, more general, makes less ad-hoc assumptions when the experimental conditions are changed slightly, and leads to more precise predictions concerning statistics of empirical data.

So much for an example of an early information-processing theory of inductive learning. The Gregg and Simon model can be seen as an predecessor of more explicit problem-solving models of induction as developed in Simon & Lea (1974), for example.

2.2.2 Constraints on Induction

The main problem of induction has been stated by the 19th-century American philosopher C. S. Peirce (1931-1958): Given the multitude of hypotheses that can be based on a particular set of observations, how is it that science makes any progress? (see also Keil, 1981; Holland et al., 1986). In a more general context, this is the question of constraints on inductive reasoning. Peirce saw only one solution to this problem: man must be equipped with apriori, ultimately biologically founded biases for the induction task. Without such predispositions, knowledge cannot be acquired. In this century, it was Chomsky who revived the discussion of the role of constraints on induction. Without a number of inborn constraints, so goes Chomsky's (1965) argument, it would be impossible for a child to acquire its first language.

Constraints on induction must not necessarily be innate. To state it dichotomically, we may think of constraints as either to be of a very general kind, or to be strongly context-dependent. For example, Keil (1981) takes the position that constraint on inductive knowledge acquisition are domain-specific, providing support from observations on cognitive development. For instance, different sets of constraints are needed to describe developmental trends in acquiring number concepts and deductive reasoning. It is not really satisfying to assume that constraints on induction can only be described in domain-specific terms. Scientists want to find universal constraints, invariants on inductive reasoning that are as general as possible.

In order to find more general constraints on induction, we have to reconsider the way constraints are expressed. Keil (1981) anchors his analysis in a Chomskyean (1965) definition where constraints are "formal restrictions that limit the class of logically possible knowledge structures that can normally be used in a given domain" (Keil, 1981, p. 198). Thus, constraints are defined in terms of structures, in terms of characteristics of learning *products*. It may now actually be the case that constraints formulated in terms of characteristics of the specific product of induction have to be context-dependent. In order to find domain-independent constraints, we might do better not to look at the products, but at the *process* of induction, i.e., to look for *process constraints* (Medin, Wattenmaker & Michalski, 1987). Product constraints will only yield a coherent picture if many different learning processes result in the same output, or if the domain limits the possible processing mechanisms. By focusing on processing principles, on the other hand, variations in performance can be explained with a few processing mechanisms.

To illustrate, constraints on human induction have almost exclusively been formulated in terms of characteristics of the products of induction: Humans prefer in some sense 'simple' rules over complex ones (Neisser & Weene, 1962); they prefer conjunctive rather than disjunctive rules (Mervis & Rosch, 1981); there is a preference for positive over negative features (Wason & Johnson-Laird, 1972). However, these constraints (for more see Medin et al., 1987; Medin & Smith, 1984) do neither hold universally - there are always exceptions - not are they always compatible with each other. In the realm of the psychology of scientific reasoning, the most famous product constraint is probably the formulation of a *confirmation bias*: that (untrained, but often also trained) scientists look for confirmatory instead of disconfirmatory evidence when testing a hypothesis or theory (Wason, 1960). However, this tendency is very sensitive to context factors; for example, it has been found that when people test every-day hypotheses, they are looking for disconfirmatory evidence (e.g., Johnson-Laird, Legrenzi & Legrenzi, 1972). This points to the possibility that people to not prefer confirmatory evidence, where 'confirmatory' is defined in terms of the logical relation between hypothesis and evidence, but that we have to look for information processing mechanisms that result in such a preference.

The distinction between product and process constraints is most precisely captured in machine learning programs that perform induction. In order to constrain the kind and number of inductive hypotheses, such systems incorporate biases formulated as constraints on the products of induction (implemented as evaluation functions, filter functions) and/or biases realized in form of processing constraints, that is, in form of operators used to construct generalizations and in the way application of these operators is controlled (see Michalski, 1983, for an overview, and Utgoff, 1986). In the models developed by myself (Chapter 4, 6), process biases are employed.

The problem solving approach provides a framework wherein induction and discovery are subject to constraints that are derived from the general nature of information-processing systems which interact with their environment by pursuing goals and receiving feedback about their success in obtaining the goals. Within this framework, tentative answers to Peirce's crucial question are: (a) inductive inference is triggered by a system's prediction and explanation failures, that is, when its prediction about the state of the environment turns out to be wrong, or when an unusual event is observed that cannot be explained by the system; (b) the amount and content of inductive inferencing is be limited by the system's current goals. These two constraints--being goal-related and failure-triggered--ensure that the inferences will be plausible and relevant to the system's goals and therefore help to improve its performance. This 'pragmatic' approach to induction (Holland et al., 1986) stays in contrast with the 'syntactic' perspective that has dominated philosophical and psychological considerations on inductive reasoning. The syntactic approach ignores the goals and the problem-solving context of the inducing system and focuses exclusively on the structural, or logical features of the stimuli being processed. Concept identification research in psychology and artificial intelligence, where materials are chosen that deliberately have no relevance for the system (such as the *thin red triangle*) is a paradigmatic example for the syntactic approach. Realistic studies of induction, however, have to be placed within a pragmatic framework:

> In our conception of the kind of processing system that can place induction in a pragmatic context, the central assumptions are that induction is (a) directed by problem-solving activity and (b) based on feedback regarding the success and failure of predictions generated by the system (Holland et al., 1986, p. 9).

2.3 Production Systems As Tools For Modeling Problem Solving and Learning

Cognitive actions are most often represented in form of *production rules*, a formalism introduced for purposes of cognitive simulation by Newell & Simon (1972) originally developed by Post (1943) in mathematical logic. A production rule (or, short, *production*) has two parts, a *condition side* and an *action side*. The condition side specifies a pattern of information that the system can recognize. The action side holds the description of an action that can be executed when the condition part is satisfied. A production constitutes therefore a single *IF...THEN...* decision. The pattern in the condition side of a production can refer to information about the external situation, to the internal state of the problem solver (usually expressed as his current 'goals'), and to information stored in long-term memory. A production might thus look like this (paraphrased in English):

```
IF      (Goal = Solve Tower Of Hanoi Problem)
        (Disc-A on Peg-1)
        (Disc-B on Peg-1)
        (Empty Peg-2)
THEN    (Move Disc-B to Peg-2)
```

A cognitive simulation model that is written as a set of production rules is called a *production system*. Production systems are executed by means of a production system *interpreter*. In its most basic form, a production system interpreter works with two interacting data structures: (1) A *working memory* holding a set of symbols called working memory *elements*; (2) a *production memory* consisting of productions (Figure 2.1). The productions are brought into contact with the data from working memory by means of the *recognize-act* cycle, which consists of three stages (Figure 2.2):

- The *match* process, which finds productions whose conditions match against the current state of working memory; the same rule may match against memory in different ways, and each such mapping is called an *instantiation*.

- The *conflict resolution* process, which selects on or more of the instantiated productions for application.

- The *act* process, which applies the instantiated actions of the selected rules, thus modifying the content of working memory (Neches et al., 1987).

This cycle stops when no more rule can be applied or when an explicit halt command is encountered.

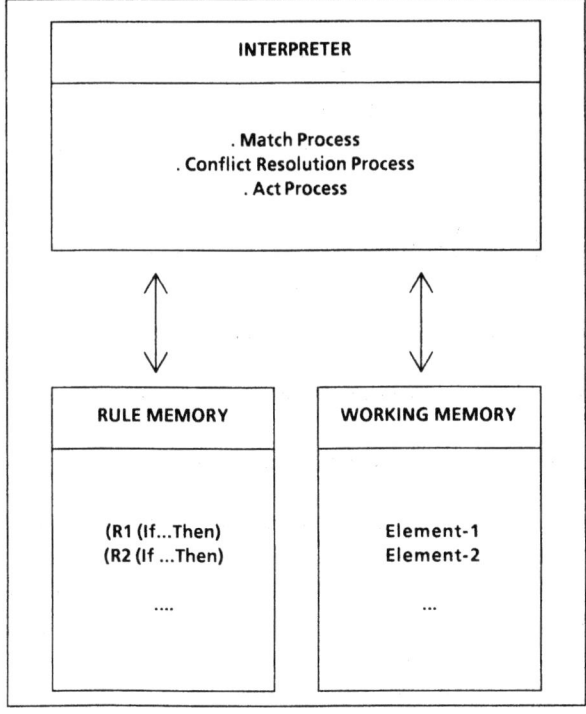

Fig. 2.1: Main components of a production system interpreter

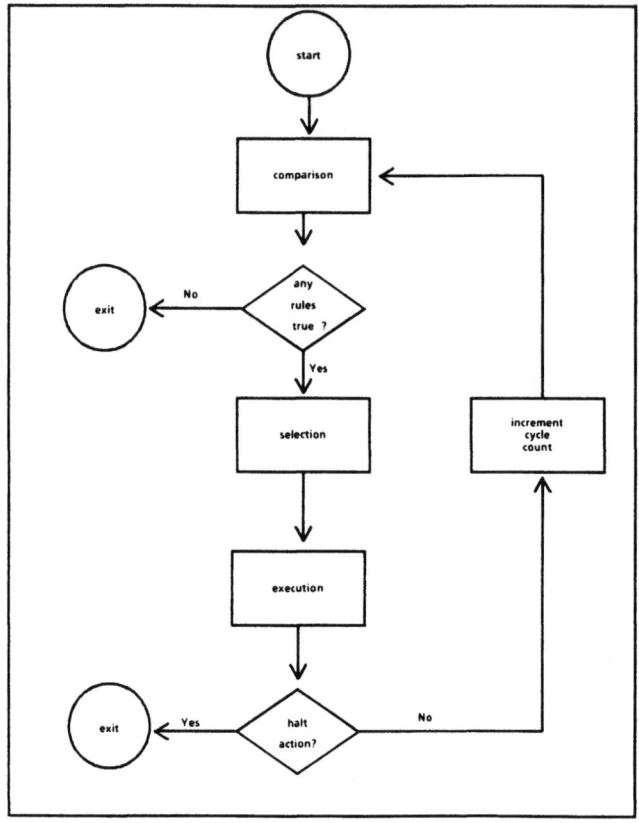

Fig.2.2: Control structure of a production system interpreter (after Ohlsson & Langley, 1986)

2.3.1 Production Systems as Theories of Cognitive Architecture

Seen as a formal notation, production systems are equivalent to an universal turing machine (Anderson, 1976) and hence could be replaced by any other formal language which has that power. Why then write cognitive simulation programs in form of production systems - which are notoriously difficult to code and understand - and not, say, in Lisp or Pascal? The point is that we want to impose a very specific meaning on a production system interpreter by seeing it as a theory about the architecture of the human information processing system. This interpretation was originally put forward by Newell (1967) and is elaborated in Newell & Simon (1972) and Anderson (1976, 1983). The notion of a cognitive architecture refers to the *invariant* features of the human information processing system. By building cognitive simulation models using the production system framework, many assumptions are made about the nature of this architecture. Newell & Simon (1972) claim psychological plausibility for the production system formalism based on the following arguments:

1. *Homogeneity.* Each rule has the condition-action structure and carries approximately the same amount of information.
2. *Independence.* A production is comparatively independent from other rules in the system. This makes is easy to add new rules or remove old ones. In particular, the rule set can be modified at run time and thus allows to model the incremental learning processes typical for humans.
3. *Parallel/serial nature.* Pattern recognition in production systems is carried out in parallel (in principle), rule application serially. This seems to be typical for the corresponding processes in human problem solvers.
4. *Stimulus-response flavor.* Many of the observations made within the framework of behavioristic learning theories can be modeled with production system models, with the additional advantage that hypotheses about mental processes can be made explicit and can be related to performance.
5. *Goal-driven behavior.* Besides accounting for the data-driven aspects of human behavior, goal-driven information processing can easily be incorporated.
6. *Modeling Memory.* The relation between the working memory data structure as a model of human short-term memory and production memory as an analog to long-term memory are obvious. The matching and conflict resolution process resembles principles of retrieval and focus of attention.

Even though the production system framework introduces significant constraints on the kinds of computation going on in the human mind (mainly due to its pattern matching character), production system architectures can vary along various dimensions, thus introducing considerable degrees of freedom. Neches et al. (1987) mention the following issues:

- *Working memory issues.* Is there a single working memory, or multiple specialized ones (e.g., one for data, one for goals)? What is the basic form of working memory elements? Is there decay and forgetting? Can forgotten elements be retrieved, for example, by spreading activation?

- *Production Memory Issues.* Is there a single production memory, or multiple specialized ones? Do productions have quantitative attributes attached such as a strength value? What is the expressive power of conditions? For instance, can arbitrary conditions be included? Can sets be matched? What is the expressive power of actions? For example, can arbitrary functions be evoked? Finally, what is the nature of the match process? Is partial matching possible? Does the matcher find all matched rules?

- *Conflict Resolution Issues.* How does the interpreter order instantiations of productions? How does it select instantiations based on this ordering? For example, does it select only the best instantiation, or all above a threshold value?

2.3.2 Rule Acquisition in Production Systems

One feature of production systems is essential for modeling (discovery) learning processes: they can be self-modifying. Mechanisms have been developed so that a production rule can insert new productions during run-time into production memory and consider them immediately for matching (*rule designation*). Further learning mechanisms allow a production system to modify quantitative attributes of its rules (e.g., *strengthening*). And there are techniques to automatically construct new rules by processes such as *generalization, discrimination, proceduralization*, and rule *composition*.

With respect to self-modification, a production system programmer must make decisions on the following points: (a) What should the basic learning mechanisms be? For example, generalization, discrimination, proceduralization, strengthening; (b) When is learning evoked? Whenever the system makes an error each time a rule is applied? Finally, (c) How do the learning mechanisms interact? For instance, will they compete or work cooperatively?

In general, learning can affect a production system at three different points (Neches et al., 1987): at the match process, during conflict resolution, and at production application. With respect to each of these points, different learning mechanisms have been proposed.

Production Matching. Adding new productions to the set of existing rules is the most obvious way of influencing the matching process. This requires a *build* or *designation* facility (Waterman, 1975) to create new rules which are inserted as a whole. Another way to add productions is to modify the condition part of existing rules by means of *generalization* and *discrimination* mechanisms (Anderson, Kline, & Beasley, 1980). The two main ways to form more general or more specific rules in production systems are to add conditions, thus generating more specific rules, or to delete conditions and thereby making them more general; or to replace variables with constant terms, or vice versa.

Conflict Resolution. Selecting from a set of matching rules is a particularly important aspect since it allows to realize the incremental nature of most learning processes: New productions are treated as guesses about the correct knowledge and the learning system has to keep a balance between trying out new productions in order to improve its performance, and performing stably by relying on established productions. In order to maintain this balance, earlier conflict resolution schemes that relied on a fixed ordering of the rules (as in PSG (Newell & McDermott, 1975), or in PAS (Waterman, 1975)) have become replaced by schemes that incorporate various weight and strength parameters (Anderson, 1976; Langley, 1987).

Production Execution. Obviously, a rule based system can change its behavior by learning about new domain operators, that is, by discovering new action sides of rules. Also, automatic learning mechanisms to modify both the condition side and the action side of a rule as a result of a rule's history were suggested: *composition* (Lewis, 1978), *chunking* (Newell & Rosenbloom, 1981), and *proceduralization* (Neves & Anderson, 1981). All these methods do not really create new operators, but perform recombinations of existing rules into new rules. These forms of learning depend on numerous applications of a rule.

2.4 Hypothesis Testing and Evidence Evaluation: Experimental Research

The process of scientific discovery is usually conceived as a cycle with three components. Theories are formulated and predictions are derived from them, then data are gathered bearing on the theory, and, finally, theories are tested by comparing the predictions with the data. Laboratory studies in psychology on inductive reasoning and scientific reasoning in particular have little to say on the first component, how theories (or, less ambitious, hypotheses) are created. These studies are mostly concerned with how hypotheses are tested.

2.4.1 Hypothesis Testing

One of the classical psychological research paradigms, concept identification (Bruner et al., 1956; Levine, 1966; Bower & Trabasso, 1964), is also a paradigmatic instance of research on hypothesis testing. In this paradigm, the experimenter selects a class of objects, for example, visual stimuli with different shape, color and thickness. The experimenter further decides on a subset of the objects to belong to a target class, the *concept*; for instance, the concept may be the class of all *round and red* objects. It is the task of the subject to find out the classification rule that defines the concept. For this, the subject is either confronted with stimuli selected by the experimenter (*reception* paradigm) or can select stimuli for himself (*selection* paradigm). In any case, he has to decide for each stimulus whether he thinks it belongs to the class or not. The experimenter responds with saying *yes*, the stimulus is an instance of the class, or *no*, it is not. The subject is said to have identified the concept when he classifies the stimuli correctly.

Directly derived from the concept formation paradigm is a problem type frequently used in psychological studies on the scientific reasoning process: the "2-4-6" rule-discovery task invented by Wason (1960, 1968). In this task, subjects are shown a sequence of three numbers (such as 2-4-6) and are asked to discover the rule that covers the example sequence. They can generate whatever number triple they want and ask the experimenter whether that sequence belongs to the target set or not; the experimenter responded with *yes* (fits the rule) or *no* (does not fit). Typically, subjects generate quickly a plausible hypothesis, one much more complex than the one the experimenter had in mind; for example, *three consecutive even numbers*. Further, subjects almost exclusively look for confirmatory instead of disconfirmatory evidence: They generate mostly number triples they believe would belong to the target set than triples they thought would not belong to it (these two possibilities are explained in the experiment instructions). This finding initiated a lot of research on the so called *confirmation bias* of human inductive thinking (e.g., Wason, 1960; Wetherick, 1962; Gorman & Gorman, 1984). This bias refers to peoples' failure to (a) seek disconfirmatory evidence, (b) utilize disconfirmatory evidence when it is available, (c) test alternative hypotheses, and (d) consider whether evidence supporting a favored hypothesis supports alternative hypotheses as well (Tweney et al., 1981). In all the studies regarding confirmation bias, the question how people come to their hypotheses, what the content of their hypotheses is and how precisely they switch to another hypothesis is not addressed. The only concern of the analysis is with the relation between a hypothesis stated and the kind of evidence selected to test a hypothesis. I do not want to dwell further into the question whether the confirmation bias is a bias at all, i.e., a deviation from the correct mode of inferencing (for a lucid discussion of this point see Klayman & Ha, 1987). Instead, I describe another study where the preference of positive evidence is at issue, this time in a setting more akin to the discovery task studied by myself.

Mynatt, Doherty and Tweney (1978) placed subjects in a simulated research environment ('particle chamber') consisting of a computer screen with objects of different shape. Subjects were told that they had to find out about the influence of the objects on the movement of a 'particle' which they could 'shoot' into the simulated chamber. It was analyzed how subjects tested hypotheses about the laws governing the particle motion in the simulated universe. As an experimental variation, half of the subjects were instructed to behave according to Platt's *rule of strong inference*, that is, to state and test alternative hypotheses whenever possible. The data where analyzed according to the relation between confirmation/disconfirmation of the hypotheses stated and the experiments ran by the subjects. The results revealed a confirmation bias and replicated a finding from a former study with the same research paradigm (Mynatt, Doherty, & Tweney, 1977). The instruction for the experimental group to use the method of strong inference did have no effects. The authors explain this resistance to use a disconfirmatory strategy with an information processing argument: "When still groping for a means for dealing with open-ended inference problems, a disconfirmation strategy may simply overload the cognitive capacity of most people". It is a goal of my thesis to say precisely what "cognitive capacity" means and what precisely its influence on induction is. I claim that such an elaboration cannot be done without considering the content of hypotheses and of evidence, and that it has to be done using computer simulation tools in order to make tractable the interaction between certain architectural limitations (such as capacity), discovery skills and hypotheses entertained.

2.4.2 Evidence Evaluation

Evidence evaluation concerns the question how people use evidence to evaluate a theory and how evidence is used to modify a theory. In actual scientific problem solving, experimental results will have effects on the revision of hypotheses; but the currently hold hypothesis will also influence how experimental results are interpreted. Psychological research has focused almost exclusively on the first line of inference, from the experimental results to the revision of hypotheses. For example, studies on induction focus on the question how people abstract a general rule given a set of particular instances (e. g., Bruner et al., 1956; Wason, 1968). Research on causal reasoning analyzes which kind of explanations people develop for certain events and whether their causal explanations are warranted given the evidence available (Shultz & Kestenbaum, 1985; Wason &

Johnson-Laird, 1972). One prominent finding in this context is *conceptual perseveration bias* which means that people tend to retain hypothesis even if they face inconsistent evidence.

The focus of this thesis is in some sense on evidence evaluation. I'm mainly interested in how people utilize information about specific experiments to build hypotheses about general regularities of the domain. However, two differences to the usual research on evidence evaluation are important: In my discovery learning environment, learners construct the information that serves as the basis for induction on their own, in a manner interwoven with hypothesis formation; they are not simply confronted with a body of evidence that has certain structural properties controlled by an experimenter. A second distinction is that I'm not primarily concerned with the logical relation between evidence and theory (hypotheses), but with how the *content* of students' theories relates to the evidence they produced. I claim that one cannot draw a strong line between hypothesis generation on the one side, experiment construction and theory evaluation on the other. In particular, in my learning paradigm 'evidence evaluation' does not mean to deciding whether evidence bears on a theoretical statement or not (this might be a narrow but precise definition of evidence evaluation); rather, it amounts to comparing the content of feedback information to the content of a prediction and to modify the hypothesis behind the prediction as a consequence of this comparison. In this sense, hypothesis generation is understood as hypothesis *modification* and is an integral part of evidence evaluation.

2.5 Hypothesis Generation and Testing as Problem Solving

The decision to describe discovery learning as problem solving forces us to take a constructive approach, to specify computational mechanisms that can actually solve the induction problem. Such a constructive approach is not often employed in psychological research, but is the rule in another branch of the cognitive sciences, artificial intelligence research. It is therefore not surprising that many ideas that relate to models of scientific discovery have first or in parallel been conceived in the subdiscipline of artificial intelligence concerned with learning: machine learning (Michalski, Carbonell & Mitchell, 1983, 1986). Famous examples of artificial discovery systems are AM (Lenat, 1982), a program that developed a variety of concepts in mathematical number theory; Dendral (Lindsay, Buchanan & Lederberg, 1980), a system which induced molecular structure from data in the form of mass spectra; and the BACON program family (Langley, Simon, Bradshaw & Zytkow, 1987).

2.5.1 Automated Scientific Discovery

BACON is a series a computer programs that "seeks to investigate the psychology of the discovery process, and to provide an empirically tested theory of the information-processing mechanisms that are implicated in that process" (Langley et al., 1987, p.4). The programs are intended to describe sufficient mechanisms to account for discovery, not to mimic in detail human performance. Langley (1979a) developed the programs BACON.1, 2 and 3. The latest version of the system is BACON.6 as described in Langley, Zytkow, Simon & Bradshaw (1986).

The core of BACON.1's intelligence lies in a set of operators that perform tests to search for patterns in data and create new terms when such patterns are found. For example, when presented with the following instances:

	Features		
Instances	Planet	p	d
I1	Mercury	1	1
I2	Venus	8	4
I3	Earth	27	9

an operator that test for monotonic increase between two variables is triggered and creates a new term and corresponding values:

	Features			
Instances	Planet	p	d	d/p
I1	Mercury	1	1	1.0
I2	Venus	8	4	.5
I3	Earth	27	9	.33

This process continues until a term is found that is constant over all observations. In the example, this is:

	Features				
Instances	Planet	p	d	d/p	d^3/p^2
I1	Mercury	1	1	1.0	1.0
I2	Venus	8	4	.5	1.0
I3	Earth	27	9	.33	1.0

BACON.1 has operators for constancy detection, slope and intercept term creation, product creation, quotient creation, and modulo-n term creation. Furthermore, unlike later versions, BACON.1 is equipped with an operator to perform specialization. This operator is triggered when a hypothesis is contradicted by data. It specializes the hypothesis by adding a conjunctive condition. Its successor, BACON.2, can solve a large class of sequence extrapolation tasks. BACON.3 has the additional ability to create new data in the process of evaluating its hypotheses. Starting with BACON.3, it is assumed that all data are known to the system before it begins generating hypotheses,[2] so that methods to specialize previous hypotheses are not longer needed.

---

[2] Actually, the system does not know all data in advance. Rather, its operators assume that the data will result from a factorial design where measurements from each cell are available to the system. BACON.3 to BACON.6 begin generating

This feature makes the later BACON systems less relevant for modeling discovery in REFRACT, since learning there is based on incremental data acquisition: hypotheses have to be constructed before the outcomes of many experiments are known.

I will rely on some ideas developed in the context of BACON.1, but make the more pessimistic assumption that not all human learners are equipped with sophisticated data analysis knowledge as it is present in that program. In particular, I will develop a model for inducing equations that does not assume that experimental data are analyzed comprehensively before equations are derived. More generally, my goal is to develop models of scientific discovery which do resemble to some degree the performance of 'novice' discoverers, not - as in BACON - the competence of an expert (Baconian) scientist. Since BACON1.'s condition finding methods are somewhat outdated, I do not describe them here. Before I turn to discuss more recent methods for inducing conditions, I want to introduce another machine learning system that induces equations from numerical input.

ABACUS (Falkenhainer & Michalski, 1986), a program for quantitative scientific discovery, is quite interesting for my purposes since it finds equations, describing numerical data and attaches symbolic descriptions to these equations describing their scope of applicability. Thus, it combines function induction and condition induction. ABACUS can handle irrelevant variables and can derive multiple equations for characterizing disjunct subsets of data.

The algorithm developed by Falkenhainer and Michalski to induce equations given numerical input combines a method to analyze features of the input data - *proportionality graph search* - with a method to heuristically evaluate newly generated terms: *suspension search*. At the heart of proportionality graph search lies the notion of *qualitative proportionality*.

> We say that x is *qualitatively proportional to y* if, for a given percentage of the events (user specifiable), the values of x rise when the values of y rise while certain specified variables are held constant. Similarly, x and y are *inversely qualitative proportional* if x decreases as y rises for a majority of the events under the same conditions (p. 374).

Based on the notion of qualitative proportionality, the following heuristic is used:

```
IF      variables x and y are related qualitative proportional,
THEN    generate a variable equal to the quotient relation between x and y
        AND generate variables equal to the difference relations between x
        and y.

IF      variables x and y are related inversely qualitative proportional,
THEN    generate a variable equal to the product relation between x and y
        AND generate variables equal to the sum relations between x and y.
```

The details of both proportionality graph search and suspension search are not important here since they rely on the assumption that all data are known before search for an equation (or

---

hypotheses after the know about the values from one level of the factorial design, and proceed then to the reading all values from the next level.

multiple equations) starts. They are therefore not applicable straightforwardly for the REFRACT induction problem. However, a weaker version of the above mentioned heuristic to find proportional relations between variables and build quantitative relations between them will be part of my simulation models. In order to attach symbolic conditions to equations ABACUS employs the A^q algorithm (Michalski, 1983). I will not describe it further, since I will use a discrimination learning algorithm.

2.5.2 Condition Finding

Most relevant for my work is machine learning research on *learning from examples*, that is, on the development of algorithms that take descriptions of specific instances as input and return symbolic generalizations (as distinguished from learning by being told or learning from practice). This task is for the most part identical to the concept formation paradigm in psychological research. For reasons that will become clear later on, systems that perform learning from examples in the form of *discrimination learning* are particularly important for modeling discovery learning in an environment like REFRACT. Discrimination learning systems start with very general rules given some first examples, and introduce more specific rules only when the general rules lead to errors in performance. More specific rule versions are created by searching for differences between situations where the general rule was applied successfully and situations where it led to an error. Examples of systems that build on a discrimination learning algorithm are Feigenbaum's (1963) EPAM model of verbal learning behavior, Hunt, Martin & Stone's (1966) simulations of concept learning, and the ACT simulation described by Anderson et al. (1980).

An alternative approach to concept learning is based on *generalizing* where the system begins with very a specific hypothesis: Only the features appearing in the first example are considered relevant. When it encounters additional examples, it removes all those features from its current generalization that do not appear in the new example. In other words, it looks for features that are held in common by a set of positive examples. This makes it difficult for generalization based concept learners to deal with noisy data and makes it impossible to find disjunctive rules. Well known examples of the generalization approach are Winston's (1970) program that learned structural concepts such as *arch*. Winston was influenced by Bruner et al.'s (1956) analysis of concept identification in humans.

A concise review of research on condition finding in machine learning and cognitive psychology can be found in Langley (1987). He compares generalization based techniques with discrimination learning approaches, favoring the second approach. A general theory of discrimination learning within the production system formalism of cognitive architecture is outlined, and examples for successful application of this technique in various learning domains are presented: concept learning, learning search strategies for puzzle problems, language acquisition, and modeling cognitive development. The various applications indicate that discrimination learning, though based on a simple idea, is generally applicable in "any domain that involves the discovery of useful conditions on rules" (p.154).

One of the few detailed descriptions of a cognitive simulation programs (and of problems with it) is found in Anderson, Kline & Beasley's (1980) paper on ACT (*Adaptive Control of Thought*), a system that combines generalization and discrimination learning within a production system architecture. ACT learns new production rules by means of designation, strengthening, generalization, and discrimination operators. It automatically creates generalizations of existing productions. The generalization algorithm implemented in ACT is rather brute force, putting clauses from two productions into correspondence by substitution of variables. This leads almost inevitably to combinatorical problems. Therefore, certain constraints are imposed on the generalizer to guarantee computational efficiency: (1) There is a limit on the amount of computing time that will be spent generalizing any pair of productions; (2) ACT only attempts to build generalizations for productions newly designated. ACT's generalization mechanism will often lead to overly general productions. One way to remove over-generalized rules is by means of discrimination, that is, by adding conditions to a production. The authors describe to some length how the complex interplay between attributes of rules such as their strength, recency, and specificity during the rule selection step contributes to performance of the system that makes it a candidate for modeling human learning. Anderson & Kline (1979) present results of using ACT in concept formation tasks.

2.5.3 Learning By Analogy

Hypotheses in a learning situation are either generated based on *prior* knowledge, i.e., on knowledge the learner brings into the experimental setting; or hypothesis generation is based mainly on knowledge acquired *within* the experimental setting. Prior (domain specific) knowledge enters the scene by means of *analogical reasoning*. When confronted with a new problem, human subjects often attempt to recall similar situations and to transfer from there to the current problem. A typical situation where people rely heavily on former experience is when they have to learn about a new technical device (Shrager, 1985; 1987; Klahr & Dunbar, 1988). The details of what triggers the analogical inference and how precisely knowledge transfer takes place are not known yet (Gick & Holyoak, 1983; Gentner, 1983; Carbonell, 1986). Analogical reasoning is most useful to transfer domain-specific knowledge or procedures from one area to another.

For the learning task studied by myself, the influence of domain-specific prior knowledge is minimized in that I assume that the discovery learner is a novice with respect to the domain. (For the subjects of the study described later, it was made sure that subjects didn't know about the relevant physics.) Influence of domain knowledge (here: optics) is further minimized in that the discovery learner does not have to recur to explanatory concepts from physics (such as force) in order to be able to solve the discovery problem. Hence, knowledge acquired in the course of learning will determine to a much greater degree whether the learner can solve the discovery task than knowledge brought to the task by analogical transfer.

2.6 Conclusions

From looking into the research literature on induction, I conclude that answers to the question of how hypotheses are generated and modified can rather be found in artificial intelligence research and in computational approaches to cognitive psychology than in experimental studies on induction and scientific discovery. The view of inductive hypothesis generation/modification as a computational search process provides us with the means to formulate precise and detailed yet tractable theories of discovery learning. I further described a specific computational formalism that is often used in cognitive modeling research: production systems. Their general relation to problem solving theory was outlined and it was argued that with this programming formalism constraints on human cognition are readily expressed. I then looked into some specific automated discovery systems, mostly developed in the context of machine learning research. In general, these models (e.g., BACON, ABACUS) are not meant as simulations of human scientists; if they are used in this way, they refer to the trained scientist, not comparable to the 'novice' discoverer that is the subject of my research. Finally, I discussed two approaches to the condition finding problem in the context of machine learning from examples: generalization and discrimination. A discrimination strategy will be part of the models developed later for discovery learning in REFRACT. But let me now turn away for a while from the question how machines learn and let me describe how humans learn, in the REFRACT discovery environment.

3. The Discovery World Program REFRACT

3.1 Introduction

In order to study the processes involved in discovery learning in a somewhat realistic context without loosing all control over students' behavior, I designed a discovery learning environment for geometrical optics: REFRACT. It is realized as an interactive simulation program that runs on a computer workstation. I will first introduce the domain, Snell's law. In the main part of the chapter, I describe how a user (student) can work with REFRACT, and which student data are recorded.

3.2 The Domain: Geometrical Optics

REFRACT is an interactive simulation program that deals with optical refraction on single surfaces (Snell's Law) and thin lenses (Lens Makers' equation). It has been designed to provide a discovery learning environment wherein the phenomenon of refraction can be explored in a self-guided way. While learning about geometrical optics, the student has the opportunity to engage in science-related activities such as conducting experiments, recording and analyzing data, and formulating and testing hypotheses. Since the program is primarily used as a research tool, it keeps detailed records of student actions.

The physics of refraction phenomena is concerned with the question of what happens to light rays when they pass from one medium (e.g., air) to another (e.g., water). In REFRACT, light rays are assumed to always travel from air to the various media of glass, flintglass, and diamond.

Two categories of experiments are available in REFRACT, depending on whether the source of light is a point or an extended object such as an illuminated arrow. In the first case, the path of single rays have to be predicted (ray experiments). In the second case, students have to predict the extended object's image (lens experiments). In the next chapter where I report observations from a study, I will restrict the analysis to ray experiments, i.e., to the question how students discover Snell's law. Therefore, the presentation of the Thin Lens law in REFRACT is not covered in this chapter. For a complete description of REFRACT's user interface see Reimann (1988a).

Figure 3.1 shows the labels for points, distances, and angles in the discovery world. The figure depicts a vertical cross-section through the center of curvature of the medium. The horizontal line is called *CenterRay*; it is a ray which is not refracted since it passes through the optical center of the medium. The *Normal* is a geometrical help line used to measure certain angles. *Theta1*, the angle of incidence, and *Theta2*, the angle of refraction, are measured with respect to the *Normal*.

Other angles are measured with respect to the *CenterRay*: (1) *Alpha*, the angle the incoming ray makes with the *CenterRay*, (2) *Beta*, the angle the *Normal* forms with the *CenterRay*, and (3) *Gamma*, the angle between the refracted ray and the *CenterRay*. Important points are *O*, the origin of the light ray; *A*, the point where the incoming ray hits the medium surface, and *I*, the point where the refracted ray intersects the *CenterRay*, also called the *Image Point*. *V* is defined as the intersection of the vertex of the medium with the *CenterRay*, and *C* is the center of curvature, i.e., a point on the x-axis located at a distance equal to the *Radius* of the curvature to the left or right of *V*.

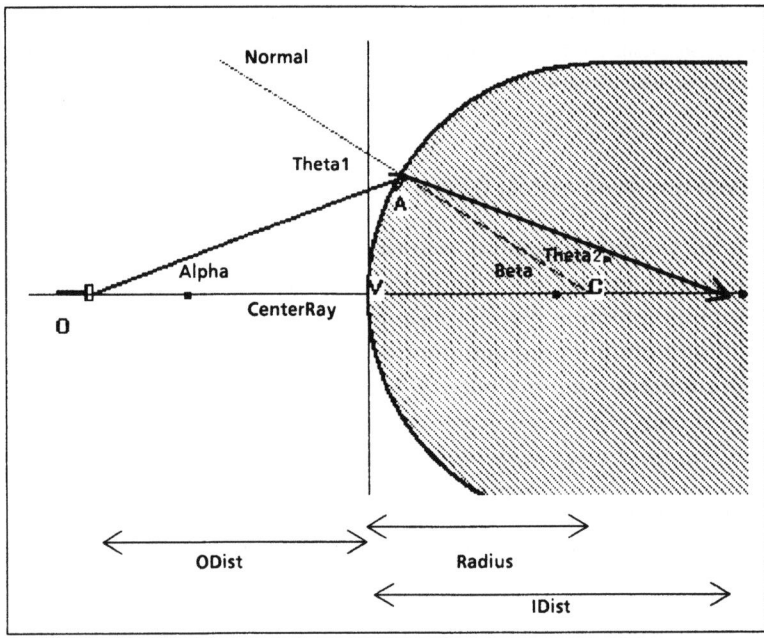

Fig. 3.1: Points, distances and angles in REFRACT

In order to make it possible for the student to discover the law of refraction and related concepts without being distracted by marginal phenomena and without having to use a lot of mathematics, the following simplifications underlie the presentation of Snell's law in REFRACT: (a) It is assumed that the light rays are ideally thin - single (laser) rays, not ray bundles; this way, only the refraction phenomenon is considered, no reflection occurs. (b) Refraction indices for the different media in REFRACT are higher than they are in reality; the angle of refraction is therefore high and rays on the screen are clearly refracted. (c) Light rays in the simulation are refracted according to an approximation to Snell's law as described below. This makes it possible for student to induce the law without having to use trigonometric calculations.

The Law Of Refraction. The Law of Refraction (Snell's Law)[1] in optics describes the phenomenon that light rays "bend" when they go from one optical medium (such as air) into another medium having a different optical density (e.g., water). This is captured in Snell's Law in terms of the *angle of incidence, Theta1,* and the *angle of refraction, Theta2:*

$$\sin \text{Theta}_1 / \sin \text{Theta}_2 = n_{21},$$

where n_{21} is a constant called the *index of refraction* of medium 2 with respect to medium 1.

The index of refraction of one medium with respect to another is defined as the ratio of the speed of light in one substance to its speed in another substance. If the index describes only one substance, it is measured with respect to vacuum. This index generally varies with wavelength. I assume in the following that wavelength is held constant at Lambda = 5890 A. Then air has an refraction index of approximately 1.0.

In REFRACT, a simplified version of Snell's Law is used. It holds in situations where the angle *Alpha*, the angle the incident ray forms with the *CenterRay*, is sufficiently small:[2]

$$\text{Theta}_1 / \text{Theta}_2 = n_{21},$$

or, to solve for the dependent variable *Theta2*, i.e., in "prediction format":

$$\text{Theta}_2 = \text{Theta}_1 / n_{21}.$$

Building on the approximative form of Snell's law, the relation between object distance *ODist* and image distance *IDist* can be expressed as:

$$n_1/\text{ODist} + n_2/\text{IDist} = (n_2 - n_1)/\text{Radius},$$

with n_1 and n_2 representing the optical densities of the two media measured with respect to vacuum, *ODist* being the object distance, and *IDist*, the image distance.

If the center of curvature *C* of a spherical medium lies on the right side of the coordinate system, the surface form is called *convex* and the *Radius* has a *positive* value; if *C* lies on the left side in the coordinate system, the surface is *concave* and the Radius *negative* (see Figure 3.2). If the radius is infinite, the medium has a *plane* surface. Note that in this case the *Normal* is a line parallel to the *CenterRay* and that *Theta1* is equal in magnitude to *Alpha*.

---

[1] Willebroad Snel van Royen (1580-1626), Dutch; called himself 'Snellius'; discovered the law in 1621.

[2] A sufficiently small value for Alpha is one less than or equal to 5 degrees (cf. Halliday & Resnick, 1986). In REFRACT, this constraint is not quite obeyed; students can select values for Alpha up to 20 degrees and the approximative equation will still hold. However, I think that the merits of enabling students to discover regularities without having to use trigonometric calculations outweigh the problems with this approximation.

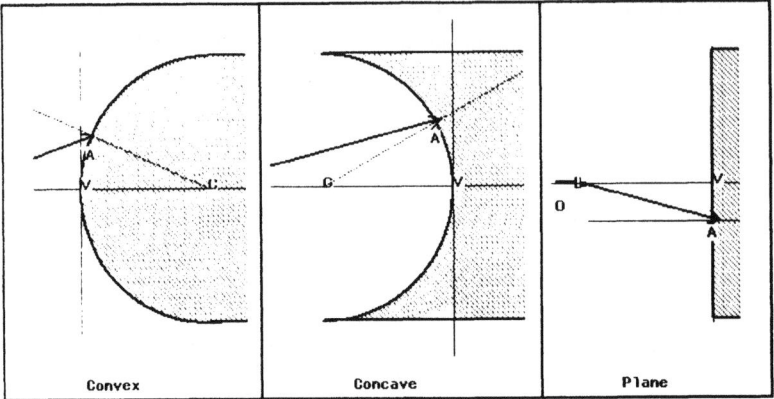

Fig. 3.2: Different surface forms

Using REFRACT, students may be expected to learn the approximative law of refraction. Since students might not generate the quantitative equations easily, a more qualitative kind of knowledge might be acquired. Examples of such insights are:

1. The refracted ray always bends towards the *Normal.* (In REFRACT, light always travels from an optically less dense into a denser medium.)
2. The image distance is affected by the object distance, medium surface form, and the medium type.
3. The angle of refraction depends on the angle of incidence (*Theta1*) and the medium's optical density.
4. The ratio of refracted angle to incident angle is a function of some property (the optical density) of the refracting medium.

3.3 Running Experiments and Stating Predictions

The essential metaphor built into REFRACT is that of a *laboratory* for optics. The student is told that his situation is similar to that of a scientist who performs experiments in optics in order to discover something about the laws in this domain. Experiments are simulated by means of a computer. In order to learn about refraction phenomena, the user can design experiments, observing what happens and storing relevant data in a computer simulated notebook. After designing an experiment and before seeing the result the student has to predict its outcome. Note that in each experiment a prediction has to be formulated (about the path of rays) before feedback can be received.

Phrased in more psychological terms, students' activities are organized according to what we call a SPFP sequence (Spada, Reimann & Häusler, 1983):

- *Search* for information by conducting experiments and making measurements;
- *Prediction* of the experimental outcome;
- *Feedback* on the outcome; and
- further *Processing* of feedback information.

SPFP sequences constitute an experimental setting that is particularly well suited for studying active discovery learning. First of all, it gives structure to the student's interaction with the system. Secondly, the SPFP sequence is a useful strategy to assess the process aspects of hypothesis generation and testing behavior. The subcomponents of this process--data selection, hypothesis/prediction formation, and hypothesis revision--are separated from each other without seriously disturbing the natural path of inductive learning. The computer can easily keep track of students' activities by recording the screen actions. Furthermore, computer records can be enriched by prompting the student for verbal comments.

The basic cycle of experimentation in REFRACT consists of three main steps: Designing an experiment, predicting its outcome, and processing feedback. During the course of the learning session, the student can conduct whatever experiments he wants to pursue, within the limitations of the program. The ultimate goal is to find out as much as possible about the laws governing the phenomena.

The key features of the implementation are:

- The program provides ample opportunity for the learner to steer the process of forming and testing hypotheses through simulated experiments.
- The phenomena the student can learn about are presented in idealized form: The numerical relations are simplified and extreme refraction indices are chosen to make the refraction phenomena visually more salient.
- The process of information selection, hypothesis formation and hypothesis testing is sequenced into separate steps, thus allowing for a detailed assessment of the student's actions.
- Predictions about the phenomena can be made on different levels of precision according to the student's knowledge about the domain.
- Information about experiments in the learning environment is given in graphical and numerical form.
- Tools are given to the student to acquire more information about one experiment and to keep track of a whole series of experiments.
- Another set of tools provide the student with the opportunity to select, organize and manipulate (numerical) information.

3.3.1 Conducting Experiments in REFRACT

Figure 3.3 depicts the main screen of the program with two central windows. In the *Laboratory Window* (abbreviated as *LabWindow*) experiments can be simulated in graphical form. The *NoteBook* window can be used to keep track of information about experiments and to manipulate

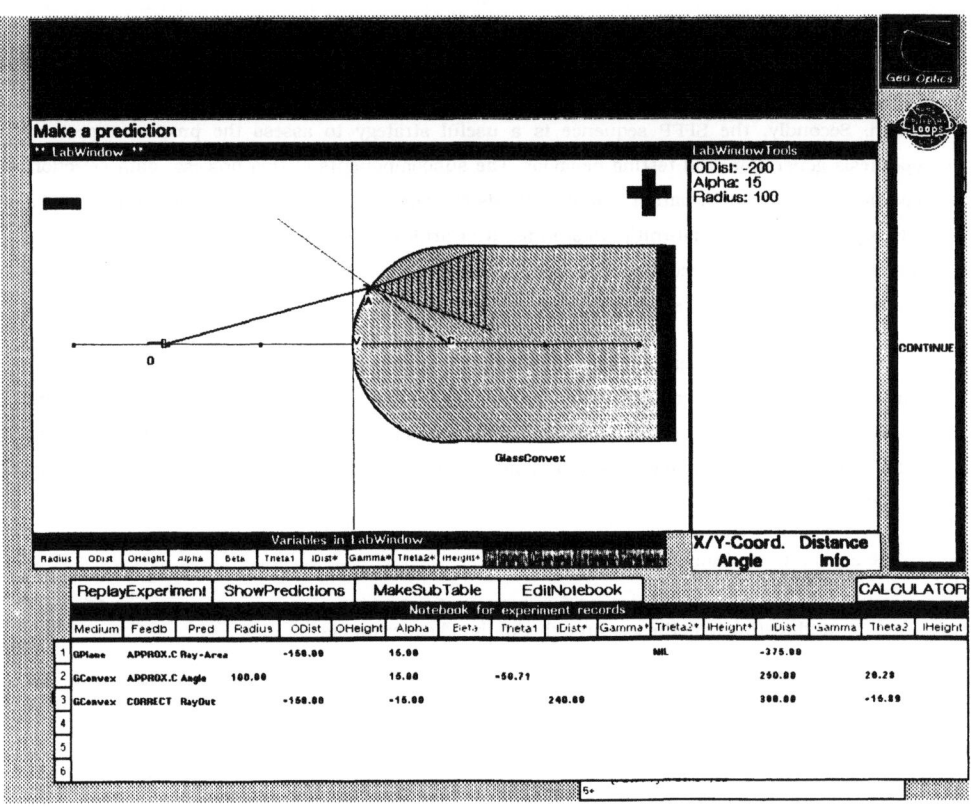

Fig.3.3: REFRACT's main screen with the *LabWindow* (in the center) and the *NoteBook* (at the bottom). The black window at the ceiling is used to display prompts to the user. The window at the right side of the *LabWindow* displays the value of variables requested by the user for the current experiment.

The *LabWindow* is connected to several supporting windows: A specific prompt window at the top of *LabWindow*, the *LabWindowTools* window at its right, and the *VariablesInLabWindow* menu at the bottom. These will be explained later. At the top of the screen is the general prompt window (shaded darkly); it is used to communicate general information about the state of the discovery world to the user.

Let us now trace the system through an experimental cycle, i.e., through the design, prediction and feedback step.

Designing an Experiment

In the design step, the student has to combine an optical medium with a ray. The user is guided through this phase by a sequence of pop-up menus. First, a medium has to be selected by choosing one of the alternatives from the *medium type menu* (Figure 3.4). The menu is ordered vertically according to surface shape and horizontally by substance. Differences in substance type are indicated as differences in the medium's degree of shading. If the student selects a convex or concave medium, there are three radii available to choose from.

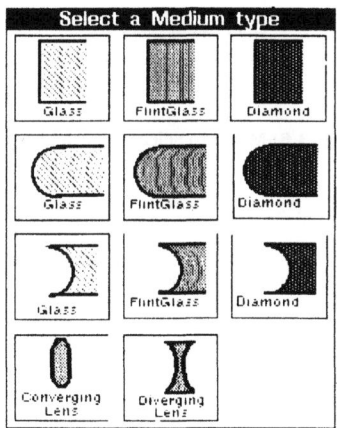

Fig.3.4: The menu used to select a medium type

Next, the user can fix the distance between the medium and the object (measured with respect to the vertex point V of the object). This is done again by selecting from a pop-up menu as displayed in Figure 3.5. The object (an icon symbolizing a "flashlight" source of a single ray) is then drawn onto the screen at the specified point. All distances and coordinates are specified in terms of screen units.

Which Distance to LightSource?
-250
-200
-150
-100
-50

Fig.3.5: Menu to select the object distance

Then the user defines the *input ray* in terms of *Alpha*, the angle it forms with the *CenterRay*. Six different values are offered (see Figure 3.6), three for rays that go upwards from 0 and three that go downwards. The program draws the input ray on the screen; also, the *Normal* at the point where the ray intersects the surface is drawn and labels are affixed for the points *O*, *A* and *V*. In Figure 3.7, a complete design of a point-source experiment is shown.

Which Input Ray?
20
15
10
-10
-15
-20

Fig. 3.6: Menu to select the angle *Alpha*

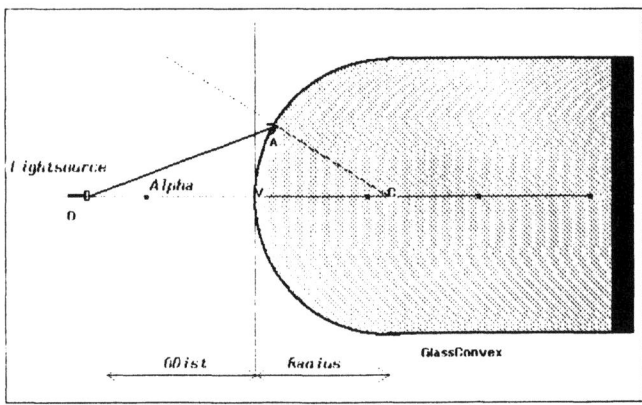

Fig.3.7: A completed design. Lables written in italics are not part of the actual display but are added here for clarity.

To summarize the design step: The student has control over four variables which comprise the independent variables of the simulation:

(1) The substance of the medium, which is varied nominally; possible values are *Glass, FlintGlass,* and *Diamond.*

(2) The shape of the mediums's surface, i.e., its radius; possible values are *plane, convex* (Radius = 100, 120, 140 screen units), or concave (Radius = -100, -120, -140).

(3) The distance from the light source, *ODist,* with five values.

(4) The angle the incoming ray forms with the *CenterRay, Alpha* (six possible values).

Note that optical density is not represented directly by the system as an independent variable, but indirectly by means of substance differences. The student is supposed to find out that there exists some property of the medium (optical density) that affects the degree to which light rays are refracted. In other words, optical density is realized in the simulation as an intervening variable, one that cannot be controlled experimentally but affects the value of dependent variables (angles of refraction and image distance).

Another design decision was to vary *Radius, ODist, Alpha* and *OHeight* discreetly instead of allowing the student to select from a continuum of values. This was done in order to make it easier for the user to repeat design selections and to speed up the assembly of a design: It is much easier for the student to construct experiments selecting values from a small menu than it is to try to control variables by drawing rays on the screen and then compute their values.

Predicting the Ray Path

In the prediction phase, the path of the refracted ray has to be specified by the student. Several options are available to perform this task. The degree of a prediction's precision can be varied according to the user's knowledge and preference. The possibility to adapt the prediction's precision has two advantages. For one, we can draw conclusions from a prediction's precision to the quality of the hypothesis which stands behind. And, secondly, we avoid frustrating the student by asking more from him than he can accomplish.

Executing a prediction is a two-step process: First a prediction type has to be selected and then the prediction has to be executed according to that particular type. Prediction types are distinguished into two groups: *graphical* and *numerical.* Graphical predictions are seen as being less precise than numerical ones. However, even within the graphical prediction types the degree of precision can be varied. The user can decide whether he wants to draw a single ray or to draw an area where the ray will presumably lie.

In the case of numerical predictions, the student is simply asked to input numbers for all the necessary aspects of a prediction, such as angle of refraction (*Theta2**), image distance (*IDist**), or image height (*IHeight**). As a convention, predicted values are distinguished from actual values by attaching a star ('*') to the former. For instance, *IDist** is the predicted image distance, *IDist* the real one. Figure 3.8 shows the possibilities for stating predictions.

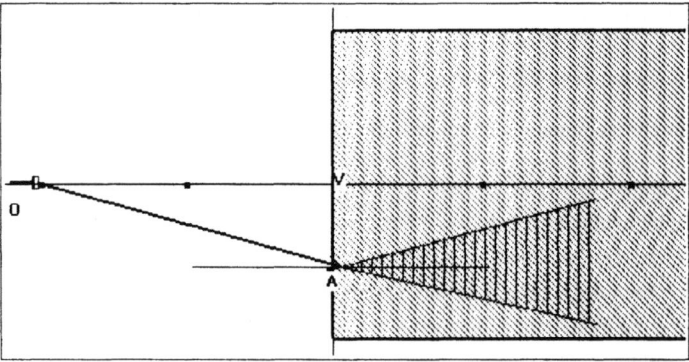

Fig.3.8a: A *RayArea* prediction. The area on the screen is specified by drawing two boundary lines strating from point A.

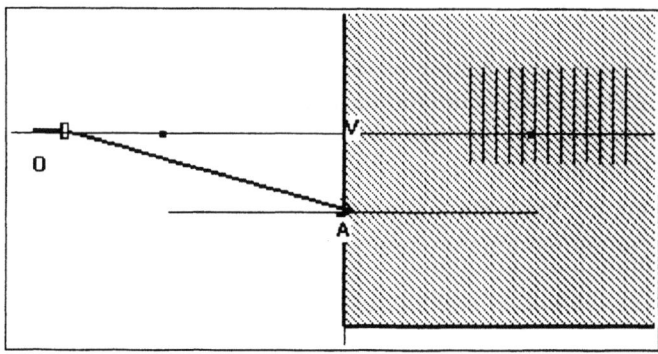

Fig.3.8b: A *ImageArea* prediction. A rectangular area is specified indicating where the image point *I* will be located.

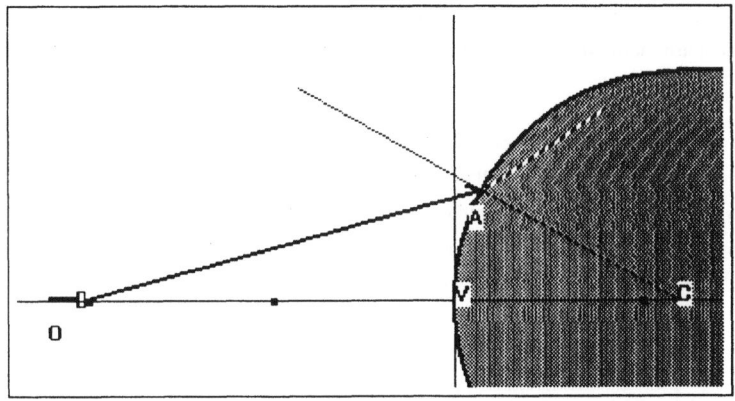

Fig.3.8c: A *Ray* prediction. The predicted ray (dashed line) is drawn onto the screen

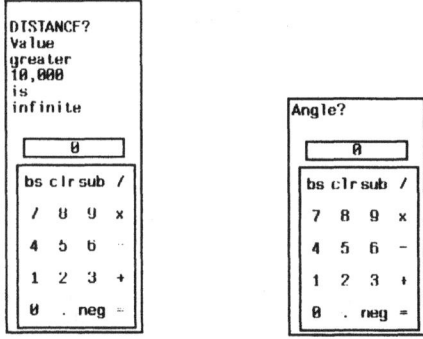

Fig.3.8d: Distance and angle prediction. The user can enter a value for the image distance *IDist* or the angle of refraction *Theta2* using the touchpads.

Receiving Feedback

Feedback is given in two forms: graphical and verbal. The correct ray is drawn into the *LabWindow*. Correct rays have arrowheads to distinguish them from predicted ones (see Figure 3.9). Verbal feedback is given in the form of a short qualification of the prediction. One of three values is assigned: *Correct*, *Approximately Correct*, or *Wrong*. This feedback is primarily provided to disambiguate the graphical feedback: Graphically small differences between predicted and correct ray path sometimes constitute major mistakes in terms of the underlying physical principles.

Verbal feedback is given according to the following criteria. If the prediction is not in a physically reasonable area, the feedback given is *wrong*; if the prediction is in the correct area, but not as precise as possible, the prediction is qualified as *approximately correct*; else it is said to be *correct*. In the case of a *ray* prediction, the approximately correct area is defined as the triangular area formed by the prolongation of the input ray through the medium and the *Normal* line.

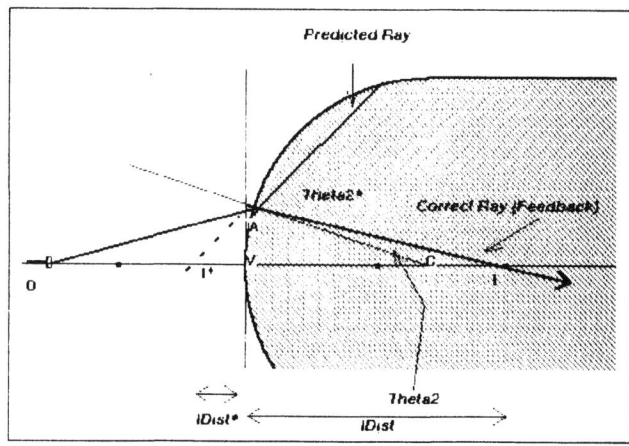

Fig.3.9: The screen after feedback is displayed in form of the correctly refracted ray (the one with an arrowhead). Labels written in italics are not part of the actual display, but added to this figure for clarity.

This concludes the description of an experimental cycle. It should be mentioned that the design and prediction step can be redone. After finishing each of these steps, the student is asked to confirm his decisions and, if not satisfied, he is put back to the beginning of the step. I will now introduce those aspects of the user interface that help the student to gather and analyze more information about experiments.

3.3.2 Gathering and Analyzing Experimental Data

Tools For Interrogating The Simulation

The user sees only graphical information about an experiment. In order to inspect numerical aspects of an experiment he can use two menus which are located at the bottom of the *LabWindow* (see Figure 3.10).

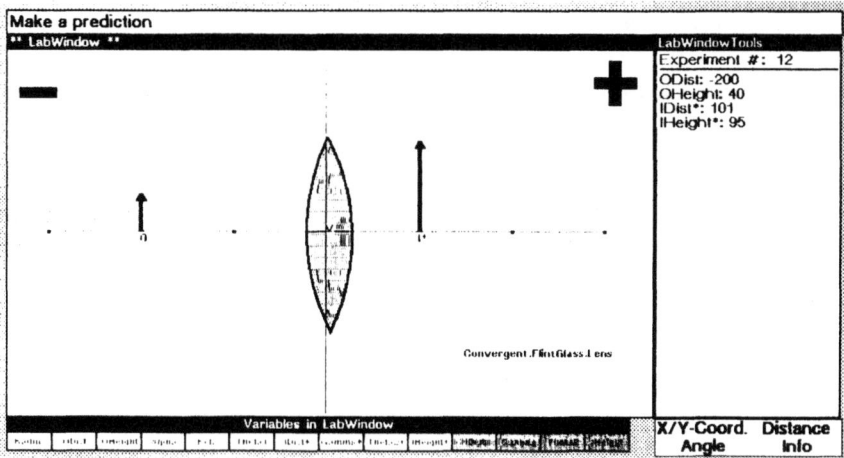

Fig.3.10: The *LabWindow* with attached inspection tools. The menus at the bottom of the *LabWindow* and to its left are sensitive to mouse operations. The items that are dark-shaded cannot be accessed yet. Values for the requested variables are printed into the window at the left of the *LabWindow*.

The *VariablesInLabWindow* menu (at the bottom of Figure 3.10) lists the names of all variables the user can access. If one of the fields in this menu is clicked with the mouse, the corresponding value will appear in the *LabWindowTools* window (attached to the right of *LabWindow*). The variables in the menu are grouped from the left to the right according to design, prediction and feedback step. Those variables which are not accessible at a certain point in an experiment are grey-shaded. For example, in Figure 3.10 the system is in the prediction step; values which pertain to the correct result are grey-shaded since they are not available yet.

If the user wants to know about numerical aspects of an experiment that are not covered by the set of prespecified variables which appear in the *VariablesInLabWindow* menu, he can make measurements anywhere in the *LabWindow* using the small menu located under the

LabWindowTools window. This menu allows the user to get coordinate values for any point in the LabWindow, to measure the distance (in the xy-direction) between any two points, and to measure the angle between any two specified lines. Furthermore, if the Info option is selected a pop-up menu appears. Selecting from this submenu will display a short definition of the chosen item.

Recording Tools: The NoteBook

The NoteBook is a tool for the user to keep track of the sequence of experiments he conducts. Further, it provides options for organizing the information in a flexible manner and performing numerical operations on experiments. Figure 3.11 shows the NoteBook with some entries. The data are organized vertically according to the sequence of experiments and horizontally according to design, prediction and feedback variables. By means of the menu at the top of the NoteBook the student can, besides other things, edit the NoteBook, or replay a previous experiment.

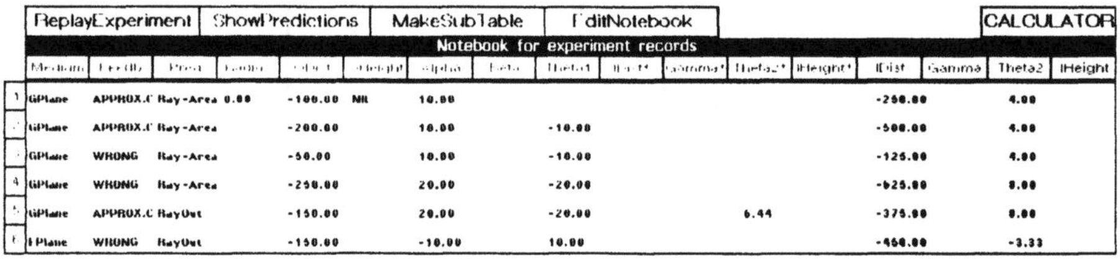

	ReplayExperiment		ShowPredictions		MakeSubTable		EditNotebook						CALCULATOR		
					Notebook for experiment records										
	Medium	FeedB	Pred	Length	x-del t	Height	alpha	beta	Theta1	th t**	Gamma	Theta2*	Height	Dist	Gamma	Theta2	Height
1	GPlane	APPROX.C	Ray-Area	0.00	-100.00	NR	10.00							-200.00		4.00	
2	GPlane	APPROX.C	Ray-Area		-200.00		10.00	-10.00						-500.00		4.00	
3	GPlane	WRONG	Ray-Area		-50.00		10.00	-10.00						-125.00		4.00	
4	GPlane	WRONG	Ray-Area		-250.00		20.00	-20.00						-625.00		8.00	
5	GPlane	APPROX.C	RayOut		-150.00		20.00	-20.00			6.44			-375.00		8.00	
6	I Plane	WRONG	RayOut		-150.00		-10.00	10.00						-450.00		-3.33	

Fig.3.11: The NoteBook. See text for explanations

The main function of the NoteBook is to display numerical aspects of experiments. This can be done by *editing* the NoteBook, i.e., by inserting values for variables into it. This is done be clicking into the field EditNotebook, selecting an experiment number from the vertical column at the *extreme* left, and selecting one or more variable names from the horizontal menu at the top of the NoteBook window. Values for more than one experiment can be inserted by selecting new experiment numbers. That means that information about experiments other than the current one may be inserted at any time. When the insert process is done, the user presses the EditNotebook field again to exit. Note that the program automatically inserts values for the first three columns of the NoteBook: Medium, Verbal Feedback, and Prediction Type used. All the other fields have to be filled in by the student.

An experiment can be *replayed* after the feedback for the current experiment is produced and before a new experiment is started. A replay is initiated by first clicking on the ReplayExperiment field of the NotebookCommand-Menu and then selecting an experiment number from the vertical number menu on the left side of the NoteBook. The selected experiment will be replayed in the LabWindow. The replay will stop after the design, prediction and feedback phase, respectively, and the user has to click the Continue bar to see the next part of the replay. In replay mode, the user can again use the *VariablesInLabWindow* menu to access numerical information.

Data Analysis Tools

In order to select, organize and manipulate information from the *NoteBook*, the user has the option to create subtables. When the *MakeSubtable* field of the *NoteBookCommand* menu is clicked, the student can select from it the variables he wants to have in a subtable. The table is then constructed, using the experiment sequence as a first sorting criterion. A maximum of 30 experiments (five times as many as in the NoteBook) can be displayed. Starting from this first table, the user has the following options: (1) He can sort the values of a variable column by pressing the left mouse button while the cursor is within the respective column; (2) he can *select values* from any column (except the experiment number and calculation variable column); (3) he can *enter an expression* that gets calculated for all experiments in the subtable by pressing the left button while the cursor is in the *CalculationColumn*. In Figure 3.12, a subtable is displayed together with a request to calculate *Theta2/Alpha* for all experiments in the table. The resulting values are displayed in a calculation column; in Figure 3.12, this is the rightmost one.

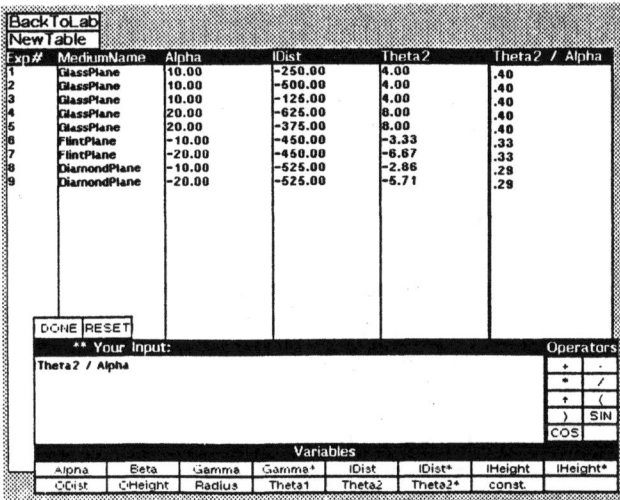

Fig.3.12: A subtable with variables *Alpha*, *IDist*, and *Theta2*. The data displayed in the subtable are based on the *NoteBook* entries. The student requested further the calculation of the term *Theta/Alpha* by means of the menu at the bottom. The resulting values are calculated for all experiments in the subtable and displayed in the rightmost column. The columns are mouse sensitive and can be used to trigger sorting-procedures.

3.4 Gathering and Recording Data from the Student

REFRACT records student actions by tracing menu selections and saving them in form of INTERLISP data structures, so called *records*. One type of record is used to store the description of an experiment (design, prediction and feedback data); on a second record type, information about how the student uses the *LabWindow* tools and the *NoteBook* is saved. Instances of both record types can be saved on file and can be read back to REFRACT. This way, a student's interaction with REFRACT can be spread over several temporal sessions, and, moreover, the data recorded can be used to analyze his behavior later on.

Besides recording many student actions automatically, REFRACT supports to some degree gathering verbal statements on a student's learning sequence. At certain points in an SPFP sequence, the *Psychologist* interrupts the student by popping up on the screen (see Figure 3.13) and states a question regarding (a) the reasons for selecting a specific design, (b) the reasons for stating a specific prediction, and (c) the insights gained from comparing prediction with feedback. The student responds to these questions verbally and his answers are tape recorded. Each of the three questions popping up on the screen is accompanied by a different sound code so that it can be recognized on the tape.

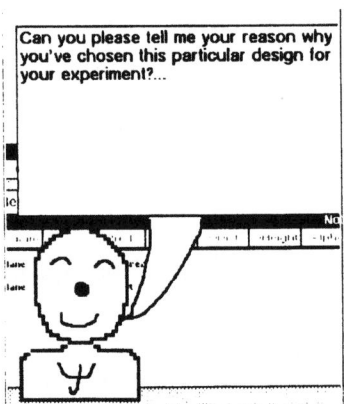

Fig.3.13: The *Psychologist Window* used to prompt the student for verbal comments

With the answers to these prompts we get additional information on students' reasoning for three crucial steps in an SPFP sequence. In particular, the chance to find out about the hypotheses behind particular predictions increases. The verbal data--note that the tape runs continuously during the session, not only when the prompting questions are displayed--combined with the other observations on students' behavior recorded automatically provide us with a fairly dense trace of data on a student's discovery behavior in REFRACT.

3.5 Conclusions

REFRACT has been described, a computer program for optics that puts students in a situation where they can learn about the domain by means of scientific discovery learning: by stating predictions, running experiments, and analyzing results. REFRACT is an environment that promotes activities which play an important part in typical scientific discovery situations. In particular, we can study the influence of hypotheses on data generation and data analysis, the processing of feedback, the interaction of domain knowledge and strategic knowledge, and the influence of problem domain representation on hypothesis generation. The student's interaction with the program is organized as an SPFP sequence, providing both structure to the student's task and a window on his hypothesis generation and testing behavior. Similar to microworld environments like the *Alternate Reality Kit* (Smith, 1986) that are implemented on machines with a bit-mapped screen, multiple windows and a mouse, REFRACT relies on *visual information* and allows for *direct manipulation* of the graphic simulations. Most of the objects appearing on the screen (ray simulations and data recording tools) can be manipulated through mouse operations. REFRACT is designed to record, structure and play back to students their behavior in the discovery world. In the current state of implementation, playing back information to the student is realized as a *replay* that frees students from the singularity and time constraints of real-world events.

The system adapts in some sense to the student's knowledge by permitting him to state predictions in several ways, ranging from purely qualitative/graphical to quantitative and precise. In order to foster the shift from a qualitative way of thinking about the domain to a more quantitative representation, REFRACT provides tools for the student to analyze experimental observations, starting from a tabular format. REFRACT is useful as a research tool since it keeps automatically track of the interaction with the user and has a facility to prompt for verbal reports.

Having described the domain and how it is presented to students, I will in the next chapter report observations on how students worked with REFRACT.

4. A Study of Learning in REFRACT

4.1 Introduction

I conducted an exploratory empirical study (Reimann, 1988b) in order to gather observations on students' learning behavior in REFRACT. These observations, seen in the light of general theoretical considerations about inductive reasoning and previous psychological research, will be later on used to guide and constrain the construction of problem-solving models of discovery learning in REFRACT, as described in Chapters 5 and 6. The presentation of the observations on discovery learning behavior of eight subjects is divided into two main parts: I begin with a summative description of the performance of all eight subjects (Section 4.3) and continue with a more detailed analysis of the hypotheses stated by two subjects (Section 4.4). The more detailed analysis of observations from the two subjects takes into account their verbal utterances. The two subjects are selected because each of them exemplifies a specific approach to discovery learning in REFRACT.

4.2 Subjects and Method

Subjects (Ss) were beginning college level students at the University of Pittsburgh with no physics exposure in the last five years. They each worked with the program for a total of about three hours, broken up into three sessions of 30 minutes, 90 minutes and 60 minutes duration, respectively. Ss were told that they were in a position similar to a scientist who wants to find out about regularities in the domain of refraction by means of experimentation; that therefore the criterion for success was not the number of their correct predictions, but the final knowledge at the end of the learning sessions. Prior to working on the computer, they were also given a guide to read that contained a short explanation of the user interface and a glossary of the terms used in the program. Ss were paid for their participation. Eight of the initially eleven Ss completed all parts.

In Section 4.3, I report observations based on data recorded automatically by the program (cf. Chapter 3). Besides presenting data for all (eight) Ss in tabular form, I will at certain points contrast the behavior of paradigmatic "good" learners (S01 and S10) with "fair" learners (S02 and S03) and with one "poor" subject (S05). This classification is based on my overall impression of subjects' performance. For the comparison between the five Ss, in addition to computer-recorded observations some verbal data are taken into account. In Section 4.4, I will go in even more detail by analyzing the verbal protocols of two subjects (S03 and S10) in order to get more process-related information on their hypothesis generation and testing behavior.

4.3 Observations From An Exploratory Study

Presentation of the results based on observations on these eight Ss is organized as follows. I begin with a qualitative characterization of Ss' final knowledge states, followed by a description of the feedback Ss received during learning, and the development of Ss' predictions. Improving predictions in REFRACT requires to identify which variables are relevant and which are not, which is in turn dependent on determining covariation between variables. In order to determine covariation effectively, Ss can use the *NoteBook* tools for analyzing data. A crucial factor influencing covariation determination is the quality of the design sequence. Finally, I report Ss' statements concerning their appreciation for the discovery learning style in REFRACT. If not explicitly mentioned otherwise, experiments dealing with lenses are ignored. I decided not to report on Ss' behavior dealing with lenses since they had considerable difficulties with this aspect, difficulties that were to some degree caused by shortcomings in the way lenses were presented in REFRACT. Further, since this is an exploratory study meant to help me generate hypotheses on discovery learning rather than to test a particular theory, I have only computed descriptive statistics.

4.3.1 Learning Effects and Final Knowledge States

Final Knowledge States

I gathered statements from five Ss which indicate what they have learned in REFRACT (Table 4.2). In order to compare how good these statements reflect what can really be learned from the microworld, important possible generalizations are gathered into Table 4.1.

Table 4.1

Important qualitative generalizations possible in REFRACT

---

1. The refracted ray bends towards the Normal. (In REFRACT, light always travels from a lighter to a denser medium.)

2. The image distance is affected by the object distance, the medium surface form, and the medium type.

3. The angle of refraction depends only on the angle of incidence (Theta1) and the medium density (substance); it is independent of the object distance and the shape of the medium.

4. The ratio of refracted angle to incident angle is a function of some property (the density) of the refracting medium. (Actually, this constant is the so called index of refraction, or the density of the refracting medium relative to the other medium's density.)

Table 4.2

Examples for final knowledge states of subjects S01, S02, S03, and S10.

	S01	The ratio of the angle of refraction and the angle of incidence is a constant and the constant is independent of the shape: *In all of them there's ratio between Theta1 and Theta2. (...) So it doesn't matter* (the shape).
		The medium's density affects the amount of refraction: *And then the flint glass and the diamond, they're almost the same in density. So they refract approximately the same amount.*
		The refracted ray bends towards the Normal: *It bends towards the Normal.*
		The image distance depends on the medium's shape: *It's going diver...It's probably going to be a virtual image* when trying a concave surface.
	S02	The angle of refraction is affected by the medium type: *With the glass plane, Theta2 was much bigger than with flint plane. ...my prediction for a diamond will be slightly less than the flint one.*
	S03	The angle of refraction depends on the medium type: *It'll probably not have much* (effect), *since I don't think the diamond had much. It had some, but this* (flint) *will be in between the plane, regular glass and diamond, I guess.*
	S10	The ratio of angle of incidence to the angle of refraction is a function of only the medium type: *There's a different relationship between Theta1 and Theta2 with the flint plane...Why would that be the case? Well, it has something to do with the flint...*

As can be seen even from this cursory evidence, Ss show considerable differences with respect to the knowledge acquired during discovery learning in REFRACT. While most of the Ss discovered that medium type effects the degree of refraction (and therefore, indirectly, that there is something like optical density), only two of the Ss considered in the table above did find out more about the quantitative nature of this effect. One subject, S05, wasn't even sure whether medium type really does make a difference.

The differences in knowledge acquired have to be seen in relation to the number of experiments ran by the different Ss. In Table 4.3, the total number of experiments (including those with lenses) that were conducted during the learning phase by each subject are displayed. Subjects ran between 23 and 31 experiments with an average of 26. The mean value when lens experiments are excluded is 17. The table shows that there is considerable variation with respect to the number of experiments and therefore difference in 'time on task'. Not considering lenses, Ss ran between 10 and 24 ray experiments.

Table 4.3

Number of experiments gor the different media

Subject	Plane	%	Sphrcl.a	%	Lenses	%	Tot.
S01	8	26.7	13	43.3	9	30.0	30
S02	6	25.0	9	37.5	9	37.5	24
S03	1	4.0	11	44.0	13	52.0	25
S05	10	43.5	6	26.1	7	30.4	23
S07	3	11.5	18	69.2	5	19.2	26
S08	4	16.7	12	50.0	8	33.3	24
S10	18	58.1	6	19.4	7	22.6	31
S11	3	10.7	7	25.0	18	64.3	28
SUM	53	25.1	82	38.9	76	36.0	211
MEAN	7		10		10		26[b]

[a]Sphrcl. = media with convex and concave surface.
[b]Mean value without lens experiments: 17.

Focusing on the frequencies for the two categories *plane* and *spherical*, we see that the two good Ss S01 and S10 spent considerable time (measured in number of experiments) in learning about one of the categories (S01 in the category *spherical*, S10 in *plane*) and a relative small amount of time in the other. S05, our poor learner, showed a similar pattern as S10, but was not able to come up with important insights for the category *plane*. The behavior of S03 is interesting in as much as for some reasons he did just one experiment in the easiest category, *plane*. Subject S02 worked on one category per experimental session (three sessions overall) and has therefore only few experiments in the *plane* category since this was the session with the shortest amount of time available to Ss for actually running experiments.

Feedback

Table 4.4 presents the distribution of feedback categories for all Ss. Since Ss were told that it would not matter whether they were wrong in a particular experiment as long as they learned something about the domain in general, these frequencies cannot directly be interpreted as indicating overall success or failure. Though not necessarily to be expected, Ss that received a substantial amount of positive feedback during the learning phase (S01, S10) also scored good in the final test and were classified as successful.

Table 4.4

Frequencies of feedback categories for plane and spherical media

Subject	C	%	A	%	W	%	Total
S01	11	52.4	3	14.3	7	33.3	21
S02	2	13.3	5	33.3	8	53.3	15
S03	0	0.0	8	66.6	4	33.3	12
S05	2	12.5	5	31.3	9	56.3	16
S07	8	38.1	11	52.4	2	9.5	21
S08	2	12.5	8	50.0	6	37.5	16
S10	9	37.5	7	29.2	8	33.3	24
S11	2	20.0	3	30.0	5	50.0	10
SUM	36	26.6	50	37.0	49	36.3	135
MEAN	4.5		6.3		6.1		16.9

Note. C - Correct, A - Approximately Correct, W - Wrong

4.3.2 Prediction Categories

In REFRACT, predictions can be made in several ways, with three degrees of precision, the least precise *area* prediction, the semi-precise *ray* prediction and the precise *number* prediction. Table 4.5 reveals how Ss stated their predictions.

Table 4.5

Distribution of prediction types

Subject	Graphic. Unprec.	%	Graphic. Precise	%	Numeric	%	Tot.
S01	5	23.8	5	23.8	11	52.4	21
S02	6	40.0	5	33.3	4	28.6	15
S03	7	58.3	5	41.7	0	0.0	12
S05	0	0.0	8	50.0	8	50.0	16
S07	4	19.1	9	42.9	8	38.1	21
S08	7	43.8	8	50.0	1	6.3	16
S10	5	20.8	6	25.0	13	54.2	24
S11	2	20.0	4	40.0	4	40.0	10
SUM	36	26.6	50	37.0	49	36.3	135
MEAN	5.9		6.3		6.1		16.9

Note. Graphic. Unprec. = RayArea predictions and ImageArea predictions; Graph. Precise = Ray predictions.

We would expect that a subject starts out with area or ray predictions as long as he is not familiar with the domain and gradually turns to more precise, i.e., quantitative predictions. From Table 4.5 we cannot gain information on the sequence of prediction type selections, but we see that certain Ss developed preferences: S01 and S10, for example, preferred the quantitative type, whereas S03 relied completely on the graphic prediction mode. In order to gain information on how Ss changed their prediction behavior during learning, the prediction type was plotted against the experiment number. Figure 4.1 shows the diagrams.

Most of the Ss did not follow a simple pattern, starting with unprecise prediction types (*area*, *ray*) and after a while switching to the precise prediction type (*angle* and *distance*). Instead, with the exception of S08, they switched between all three levels over the whole learning sequence. Some Ss, however, showed a systematic pattern. S01 and S10 adapted the prediction type to the design selection by means of the following procedure:

1. Start out with an unprecise prediction type (*area*) when working with a medium type not yet explored.
2. Look at the feedback and try to improve your prediction.
3. Improve precision on the next prediction until a satisfaction criterion is reached.
4. Goto 2.

This systematicness and their persistence in keeping medium type constant until they could improve their prediction certainly contributed towards the success of these two Ss.

To sum up: From what we know now about Ss' final knowledge state and about the form they stated their predictions in, we can draw one important conclusion: Many Ss in this study did not extensively utilize quantitative information about variables. In the extreme case (S03), quantitative information was not included at all in hypotheses. And even the best Ss (S01, S10) occasionally fell back to make predictions based on non-quantitative attributes. A possible explanation for this phenomenon is that Ss had difficulties finding out the relevant variables to base numerical predictions on, that is, had difficulties with variable selection. Let us therefore have a closer look into which variables they selected for prediction making.

4.3.3 Discrimination of Relevance

As mentioned before, many variables available in REFRACT are irrelevant for prediction purposes. For example, *Alpha*, the angle the incident ray forms with the optical center line, is not generally useful, despite the fact that it is a central independent variable, i.e., one the learner can control. *Alpha* leads to simple predictions only in case of *plane* surfaces. The second independent variable, *ODist*, the object distance (distance between the medium's vertex and the light source), is - though

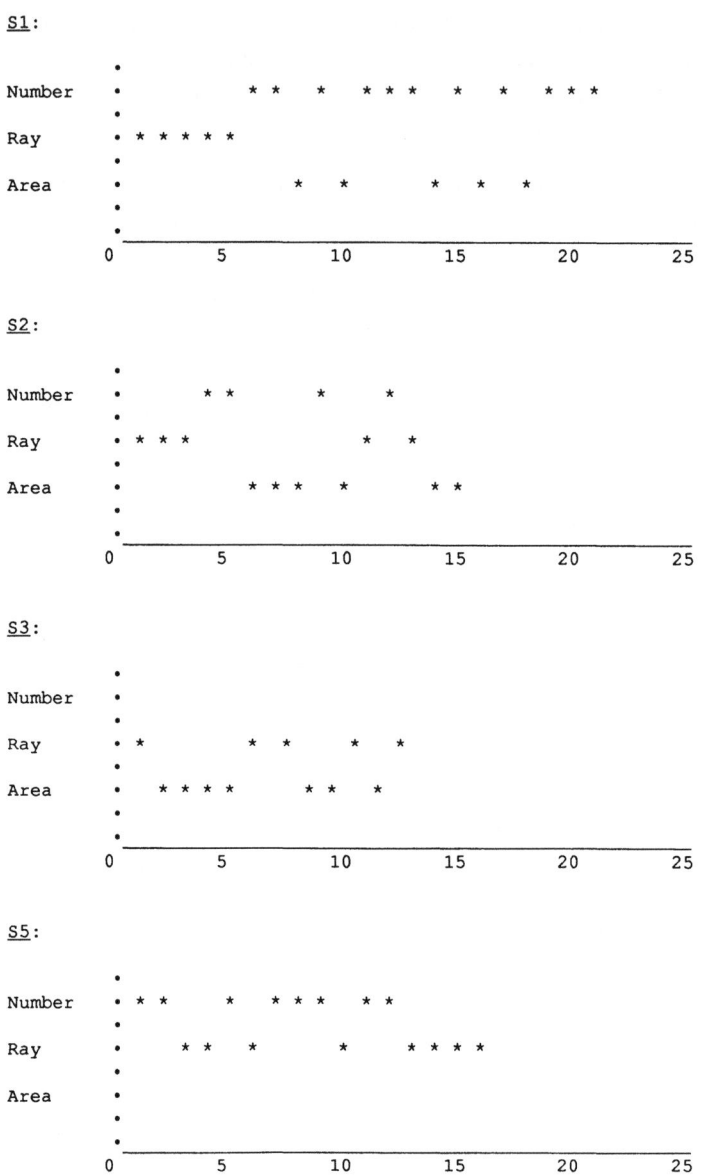

Fig. 4.1: Prediction type (Area, Ray, Number) selection over experiments

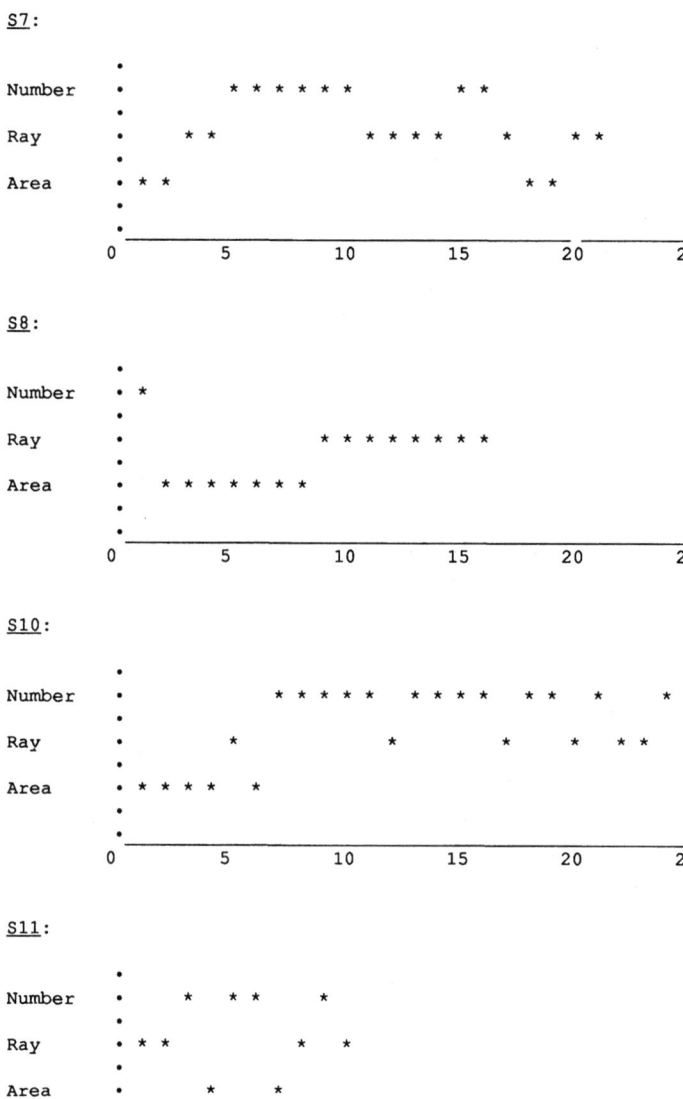

(Fig 4.1 continued)

not irrelevant - difficult to use for (numeric) predictions since the algebraic expression that relates it to the dependent variable *IDist* (image distance) is fairly complex.[1] On side of the dependent variables, *Gamma* is related to *Theta1* and *Alpha* in a simple form only for plane surfaces. In short, simple *and* general numerical relations useful for prediction making exist only between *Theta1* and *Theta2*, the angle of incidence and refraction, respectively.

S01 and S10, the better learners, figured out that *Theta1* and *Theta2* are more relevant than *Alpha* or *Gamma*. S01 successfully predicted that *Theta2/Alpha* is constant for plane surfaces (where *Alpha* and *Theta1* have the same value) but realized that this relationship is no longer true for the spherical surfaces and turned his focus to the *Thetas*. In subsequent experiments he revealed that these two are more significant. S10 came from the other side. He first found that the two *Theta* angles are relevant and then investigated the influence of other factors: "...*see what other things I can use besides the Thetas. Maybe the Alphas I should figure out first.*"

Other Ss were not as successful in discriminating relevance. S02 did not clearly realize that *Theta1* was more relevant than *Alpha*. Even during *NoteBook* recording he put in consistently values for *Alpha* but did not record the values for *Theta1*. S03 acknowledged the importance of *Theta2* since it is one of the two dependent variables (besides *IDist*) he was supposed to specify, but was unclear about *Theta1*. S05 often set for himself the goal to figure out what's relevant and what's not, but didn't have much success in doing so.

An interesting indicator in the context of discrimination of relevance is how *Theta1*, the angle of incidence, is used for prediction making. Recall that *Theta1* is the single most useful predictor. I counted how often the value of *Theta1* was fetched from the *Variables-In-LabWindow* menu during the design or prediction step of an experimental cycle, because this indicates whether a subject potentially considered this value in his prediction. Only S10, the most successful subject and the one that worked most numerically utilized *Theta1* regularly (Table 4.6).

---

[1] Recall that $Theta2 = Theta1/n_{21}$ and $n_1/ODist + n_2/IDist = (n_2-n_1) / Radius$

Table 4.6

Frequency of references to Theta1 during the design or prediction phase

Subject	Frequency
S01	2
S02	0
S03	0
S05	1
S07	1
S08	0
S10	8
S11	4

A second overall measure of how successful subjects were in discerning variables' relevance is whether they used the angle of refraction, *Theta2*, as dependent variable, or the image distance *IDist*. Only the first one stands in a simple numerical relation to an independent variable, *Theta1*. As can be seen in Table 4.7, only those Ss that concentrated on *Theta2* (and ran a significant number of experiments) were successful prediction makers. It is astonishing that S05 and S07 did not give up using *IDist* as dependent variable, despite the fact it is not straightforwardly related to any independent variable.

Table 4.7

Frequency of using Theta2 and IDist as dependent variable

Subject	Theta2	IDist
S01	11	0
S02	4	0
S03	0	1
S05	1	7
S07	1	7
S08	0	0
S10	13	0
S11	3	0

From this look at how successful Ss identified relevant IVs and DVs in REFRACT, I infer that they for the most part did not solve this problem. This shortcoming is a first reason for why Ss were not particularly good in predicting ray paths. Hypotheses based on irrelevant or only partially relevant variables led to wrong predictions. Further, since no simple relationships can be found when irrelevant or partially relevant variables are involved, Ss might have been tempted to base their predictions on pictorial features rather than on complex numerical relations.

By now we have found that Ss often did not identify promising variables. But what are the reasons for this failure? Discrimination of relevant variables must be based on observations of covariation between variables. Hence, a reason for not being able to identify relevant variables (and their relations) is that Ss were not able to find out which variables covary. Before we look at this issue, we have to analyze a prerequisite to covariance determination: design selection.

4.3.4 Design Selection

The design sequences for all Ss are provided in Appendix IV.1. For the number of experiments see Table 4.3. Overall, Ss did not construct completely factorial designs. This is, of course, not necessary: As soon as one knows that a given IV does not influence the value of the DV focused on, one can vary that variable freely, or keep it constant. However, we know already that Ss were for the most part not very good in identifying relevant variables, and we might therefore expect that varying IV values in an uncontrolled manner will make it hard for them to detect regularities.

Subjects showed significant differences with respect to the number of variables they varied in the average (Table 4.8). While S03 with an average value of 3.6 changed almost all variables between two experiments (there are four independent variables overall), S11 seldom changed more than one (average of 1.4). However, this parameter is not related to the learning outcome in the sense that the less variables are varied the better the learning outcome is. While one good subject, S10, varied very few variables at a time, the other good one, S01, altered more. This behavior has to be seen in light of Ss' knowledge about the significance of variables for prediction purposes.

Table 4.8

Average number of variable changes

Subject	Avrg. Number of Changes
S01	2.6
S02	1.5
S03	3.6
S05	2.7
S07	1.5
S08	1.7
S10	1.6
S11	1.4

In general, varying at the same time *substance* type (therefore, indirectly, optical density) and the DV that affects the IV focused on (IV *Alpha* for DV *Theta2* (and *Gamma*), IV *ODist* for DV *IDist*) makes it more difficult to derive trend information since both factors influencing the value of the DV are changed. A learner will increase his chances to find regularities in the experimental observations and the data when he keeps substance type constant and varies either *ODist* (when predicting *IDist*) or *Alpha* (when predicting *Theta2*). *ODist* may be varied freely when predicting

Theta2 (or *Alpha* for that matter), and *Alpha* may be varied freely when predicting *IDist*. Furthermore, changing from surface form *plane* to a spherical surface will change the relation between the IV *Alpha* and the DV *Theta2* (many Ss use Alpha as their first candidate for the IV since *Alpha* is under their direct control) and the relation between the IV *ODist* and the DV *IDist*.

The only subject that came close to a factorial design scheme was S11. He varied the substance type first, then the radius. The other Ss showed a less systematic behavior, either varying variables fairly haphazardly, or performing according to a scheme similar to the one outlined in the last paragraph. However, to stress the point again: Since hypothesis formation was interwoven with experiment construction, it was not necessary to construct a fully factorial design in order to perform optimally. The quality of design construction can only be judged on the basis of the knowledge the learner had. It is also interesting to note that one subject, S10, *repeated designs*. But this happened at the beginning of his learning session and it might indicate his way to get acquainted with the different options of stating predictions in REFRACT rather than a real misunderstanding of experimentation.

Subjects did not construct clearly factorial designs. This is not surprising since an interaction between knowledge state and experiment construction was expected. Still, since most of the Ss were not able to identify relevant variables early during experimentation, their habit of not controlling variables systematically should hinder them to find functional relations easily. The quality of design creation does affect the ease to determine covariation between variables. Covariation detection is the topic of the next section.

4.3.5 Determining Covariation

S01 and S10 were quick in integrating evidence from several experiments and to formulate a general hypothesis. It was harder for S02, and the other two Ss for which I looked at verbal data, S03 and in particular S05, did not succeed in this regard. Here some quotes from various subjects. (S01:) Working with plane surface: *"Theta2 and Gamma are same here. Are they always the same?" "I'm trying to see what the relationship between Gamma, Theta2 and Alpha is."* (S02:) Also while working on plane surfaces: *"With glass plane...Theta2 was much bigger than with flint plane...So my prediction for diamond one will be that it will be slightly less than the flint one."* (S05:) *"I'm trying to think of what kind of calculation I could go through to predict where this refraction is going to end up at." "I don't have an equation of anything. Did I have to derive that? ... I don't have any idea."*

In order to determine whether and how two or more variables covary, it is helpful to look over the outcomes of several experiments listed in tabular format. Furthermore, the need might arise to rearrange data for one variable in increasing or decreasing order as well as to apply mathematical operations to one or more columns of a table. REFRACT provides tools for such data analysis strategies by means of the *NoteBook* and the *Subtable* mechanisms. The question is how Ss determined covariation and whether they used these tools.

All Ss, with the exception of S05, used the *NoteBook* tool to record data from experiments (see Table 4.9). For the most part, Ss recorded design and feedback values, but did not as often enter prediction values into the *NoteBook*. This, of course, made it harder to compare prediction and feedback on a numerical level.

Table 4.9

Number of variables inserted into the NoteBook

Subject	Design	Prediction	Feedback	Total
S01	63 (59)[a]	0 (0)	44 (41)	107
S02	4 (11)	12 (32)	22 (58)	38
S03	28 (60)	4 (9)	15 (32)	47
S05	1	0	0	1
S07	27 (56)	9 (19)	12 (25)	48
S08	46 (44)	24 (23)	35 (33)	105
S10	22 (47)	9 (19)	16 (34)	47
S11	26 (50)	10 (19)	16 (31)	52
SUM	217 (49)	68 (15)	160 (36)	44
MEAN	27.1	8.5	20.0	55.6

[a] Percentage

The tools that build on the *NoteBook* were not used very often, as can be seen from Table 4.10. Note, however, that the numerical relations can be induced without using these tools, as demonstrated by S10. It could be argued that the tools were not utilized because the user interface did not support their usage. This can be only partially true, because the tools were explained and demonstrated in the first session and their usage was encouraged by the experimenter throughout all three sessions.

Table 4.10

Frequency of using NoteBook tools

Subject	Replay	Subtable Usage		Calc[c]	Total
		Create[a]	Sort[b]		
S01	4	2	2	1	9
S02	2	2	1	0	5
S03	2	2	2	1	7
S05	2	0	0	0	2
S07	5	4	3	4	16
S08	9	1	1	1	3
S10	2	1	1	1	3
S11	0	1	1	1	3
SUM	26	13	11	9	48
MEAN	3.3	1.6	1.4	1.1	6.0

[a] Create a subtable by selecting variables from the NoteBook.
[b] Sort the values in a subtable according to one variable in ascending, descending or alphabetical order.
[c] Perform calculations on one or more columns.

4.3.6 Response to Guidance-Free Learning

I expected that some students would feel comfortable with learning on their own, while others would like to have more guidance. Some statements of Ss in my small sample point to such differences. S01, for example, starts out pretty sceptical: *I'm just guessing. I have no idea. You should let us read before.* When the experimenter denies that, the subject replies: *Are you supposed to learn while you're doing this?...You're kidding.* Later into the session he feels more comfortable with the situation: *I'm getting better at this.*

S03 and S10 seemed to be quite at ease in the environment and did not make any remarks on the learning style. While I have no clue how S02 felt about it, S05 was positively upset and insecure. Towards the end of session 3, he comments: *You probably think I'm stupid. Honestly, I do well in school.*

4.3.7 Summary

Subjects in this small study showed many differences, both with respect to knowledge acquired and to learning behavior. This is not surprising, given the openness of the learning situation. While almost all Ss found out that medium substance affects refraction, only a few of them could express this in terms of functional relations between dependent and independent variables. On side of the learning process, I noticed differences with respect to control of variables, prediction making, discrimination of relevance, and determination of covariation. The later relates to Ss skill in keeping track of experimental observations and finding regularities in data. Finally, Ss felt more and less comfortable with the guidance-free learning situation.

I think that two findings are of foremost importance. First, many Ss are not able or not willing to formulate predictions which are precise in the sense that they are based on numerical relationships between variables. This is in part reflected by the fact that the tools in REFRACT which can be used to discover such relationships were not used very often. Closely related to this first point is a second central observation: the frequent use of graphical (pictorial) information. I did not expect that Ss would use this kind of information so often and constantly to build generalizations and to generate predictions. To get more information on this point and to describe Ss behavior on a level closer to computational modeling, I undertook a protocol analysis of two Ss, S03 and S10. They are particular interesting to compare since S10 is a subject that works mostly in a numerical mode, whereas S03 states his predictions often in a non-numerical fashion.

4.4 Analysis of Verbal Hypotheses

Subjects had to think out loud during their working sessions with REFRACT, and all statements were tape-recorded. The production of verbal data was partially cued by certain prompts that appeared on the screen. Ss were 'asked' by the program (1) after they assembled a design, Why they selected the design values, (2) after the prediction, Why they made the prediction, and (3) after the feedback, What they concluded from comparing prediction with feedback (see Chapter 3). Answers to theses particular statements are the main object of the investigation in this section. The goal of this analysis is to describe hypothesis modification processes and resulting knowledge structure in more detail than it would have been possible without looking at verbal reports in detail. I am particularly interested in the influence of different encodings of screen information on learning. I will compare the two subjects with respect to the verbal hypotheses they stated or that I inferred from their predictions, and with respect to some related aspects, for example, the reasons they gave for designing particular experiments. The analysis is based on a more encompassing protocol analysis described in Reimann (1988b).

4.4.1 Overview of S03's and S10's Learning Behavior

Subject S03's Hypotheses and His Discovery Behavior

The following hypotheses were inferred from S03's predictions and from his answers to the question on what his predictions are based on (Table 4.11).

Table 4.11

All hypotheses of subject S03 expressed as rules

H1	IF <any>	
	THEN Ray is bent in	
H2	IF <any>	
	THEN Ray is bent inwards slightly	
H3	IF (Surface Plane) (Substance Glass)	
	THEN Ray will go straight through	
H4	IF (Substance Diamond)	
	THEN Ray will be bent out	
H5	IF (Surface Concave)	
	THEN Ray will be bent out	
H6	IF (Surface Concave)	
	THEN Ray is bent slightly out	
H7	IF (Substance Diamond) (Surface Concave)	
	THEN Ray is bent outward and very close to the Normal.	
H8	IF (Surface convex)	
	THEN Ray is bent in	
H9	IF <any>	
	THEN As substances changes from glass to flint to diamond the refraction effect becomes smaller.	
H10	IF <any>	
	THEN As ODist increases, IDist increases	
H11	IF <any>	
	THEN As ODist increases, IDist decreases.	
H12	IF (Surface Concave) (ODist - 150)	
	THEN Ray is bent out and runs slightly under the Normal	
H13	IF (Surface Concave) (more)	
	THEN Ray is bent in and runs slightly under Normal	
H14	IF <any>	
	THEN Theta2 is affected by Alpha and ODist.	
H15	IF (ODist -50)	
	THEN Ray runs parallel with X-Axis	

Note. If identical condition parts occur in different rules, the later rule overrides its predecessor.

What is most striking is that S03 did not induce a single equation. Rather, he stated his predictions mostly in a graphical form. The corresponding hypotheses can consequently not be captured in form of equation-plus-scope, but must be described as phenomenon-description-plus-scope, where a phenomenon description is a qualitative statement about the ray path (*Ray goes up*) or a semi-quantitative relation (*As ODist increases, IDist increases*). In order to understand why this happened, we have to look into the sequence of experiments he ran and the hypotheses he derived.

E1: His first prediction is that the ray will be refracted in an upwards direction, which I take as derived from the general hypothesis, H1, that a ray is bent when it enters one of the media in REFRACT. From the feedback (cf. Figure 4.2a), he concludes that "*the correct ray didn't ... bend up as far*", an observation encoded as H2.

E2: Looking at the design of experiment 2 (Figure 4.2b), S03 states two possibilities: "*Maybe because it's plane...it will go kind of straight through, but the diamond...might have an effect that it will bend...*". He does not come to a decision and makes a prediction that covers both hypotheses. From the feedback he concludes that the second hypothesis, encoded here as H4, is correct.

E6: After having run three lens experiments, he returns to the single surface phenomena and designs an experiment with substance *glass* and surface *concave*. Note that he varies many variables in the first three ray experiments, in particular, substance type and surface form are varied together. His prediction for this experiment seems to be based on the hypothesis H5, which is like H1, H3 and H4 an example for pre-experimental knowledge: "*For some reasons I have the feeling it's going to go down*" (see Figure 4.2c).

E7: He selects the design in order to "*see what kind of effect both the diamond and the concave have together*". In trying to come up with a prediction, he looks backwards to E2 (with *concave* surface) and E6 (with *diamond* as substance). For this he has to look at the NoteBook, since the two experiments are no longer visible. It is interesting to read what he does when skimming the NoteBook: "*I look at the concave one I just did the last time. And look at the Theta2 [!]. And it was just a little, bitty angle from the Normal. (...) And it went down. The thing [the ray] was coming down and it went down [!].*" So, even when he looks at the numbers he translates them into a ray *path* description. His prediction seems to be based on H5 again, as in the last experiment with a *concave* medium.

E8: In this experiment he seems to either make a slip (it was the first experiment of his second learning session, so there was a considerable time delay since he ran E7), or he has overgeneralized H5, which was built in the context of *concave* surfaces: He predicts that the ray will be bent outwards in case of a *convex* surface (cf. Figure 4.2d), different from his very first prediction in E1.

E9: This is the only experiment where S03 makes a numerical prediction. Before predicting the image distance for this experiments, he looks into the NoteBook and compares the current design with that from E1. From that he derives H10: *As ODist increases, IDist increases*. From the feedback, he has to conclude the contrary, H11: *As ODist increases, IDist decreases*.

E10-E12: The prediction in E10 shows (cf. Figure 4.2e) that S03 so far does not consider the *Normal* as important, since he predicts (somewhat more assuredly since he uses a *ray* prediction instead of an *area* prediction) that the ray will be bent outwards, but across the *Normal*. The feedback forces him to attach an additional rule to H5, resulting in H12. However, he does not seem to be sure that this is true for other situations, since he sets up a slightly modified design (Figure 4.2f) in E11 and repeats essentially the wrong prediction from E10. In E12, he finally gets it right. Also, in this experiment, he tries to find hints in the numerical data from the NoteBook. But since his experiments are not designed systematically, he does not find anything of predictive value, only a qualitative insight that the value of *Theta2* is affected by *Alpha* and *ODist*. The following statement describes his problems probably correctly: *It's too much for my little brain.*

E16-E18: In the last three experiments on single surfaces, he does not acquire more insight into the relationships. In E16, he has problems because the small object distance lets the ray no longer be "bent in". This forces him to build yet another hypothesis covering the case of small image distances.

A discrimination based strategy seems to describe more or less correctly the way S03 modifies the scope of his hypotheses, since he uses hypotheses generated in one experiment in later experiments that differ from the creation context. While we can say nothing on how this learner constructed equations, there is evidence that he attempts to induce regularities by comparing experiments. For example, in E7 he looks for the last experiment with *concave* surface and/or substance *diamond*.

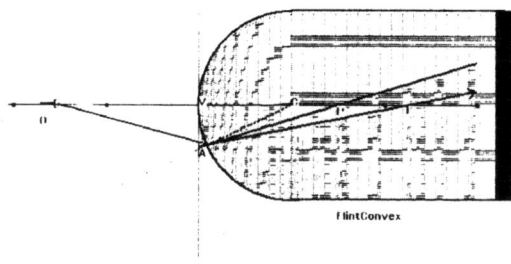

Fig. 4.2a: Experiment 1

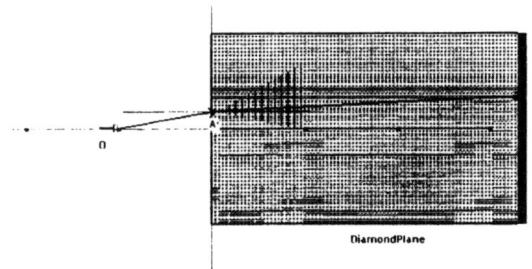

Fig. 4.2b: Experiment 2

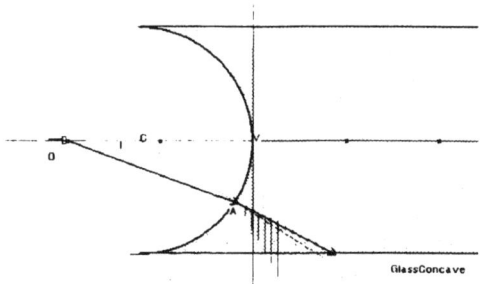

Fig. 4.2c: Experiment 6

Fig. 4.2: Graphical predictions made by Subject S03

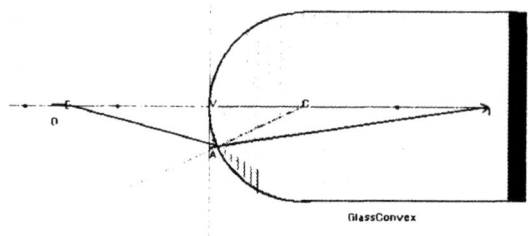

Fig. 4.2d: Experiment 8

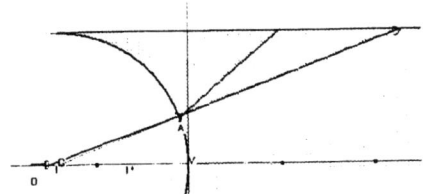

Fig. 4.2e: Experiment 10

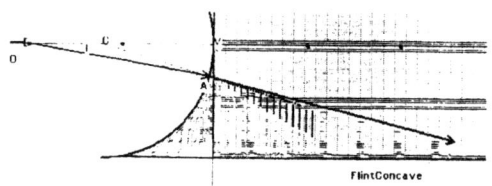

Fig. 4.2f: Experiment 11

Subject S10's Hypotheses and His Discovery Behavior

In the course of his 24 experiments with single surfaces, S10 acquired knowledge that I phrased into the rules listed in Table 4.12. Let me summarize his discovery behavior by following him through the experiment sequence.

E1 - E3. S10 starts out by repeating a design three times. Using the *RayArea* prediction form, he attempts to make the predicted area smaller and smaller and thus increasingly precise. After the third experiment of this kind in sequence he seems to realize that increasing the precision is no longer possible in this way and starts looking for numerical relations. He finds that two variables didn't change their value in the last three experiments: *Theta2* and *Gamma*. Note that he looks over three experiments column-wise, not over two experiments row-wise). S10 immediately announces that this finding is irrelevant since no independent variable was changed. Consequently, in the next experiment (E4) the design is modified: substance type is now *glass* instead of *flintglass*.

E4-E6: Since S10 dismissed the finding that *Theta2* has the same value for the first three experiments, he does not use that relation but makes again a *RayArea* prediction, which turns out to be wrong. Comparing E3 with E4, he finds that *IDist* decreased, but considers this finding as irrelevant. He also notices that the value of *Theta2* increased. However, this finding is not immediately turned into an hypothesis (or at least not in one with high enough strength), since in the next experiment, E5, a graphical prediction type is used again, a *Ray* prediction. From E4 to E5 and again to E6 he attempts to make his predictions more correct by switching prediction type from *RayArea* to *Ray*, keeping the design constant. After having run E6, S10 seems to realize--similar to the situation after E3--that improving the prediction requires a change in prediction type and begins to focus on numerical aspects. In the context of E6, he builds the hypothesis H5: IF <any design> then *Theta2* equals 4 degrees.

E7-E8: That H5 is an overgeneralization turns out in E7, where *Alpha* switches signs. S10 immediately generates a new hypothesis, H7, which incorporates the observed value for Theta2. We might hypothesize that H5 becomes discriminated here, resulting in something like H6. H7 is "tested" in E8, which is really a redundant experiment since in repeats the design from E7 needlessly.

E9-E10: From the last experiments, S10 seems to have induced another hypothesis, H8, which says that *as Alpha increases, Theta2 increases*. Since he doesn't know the exact quantitative relation yet, this leads to an approximately correct value for his prediction in E9. From E8 to E9 and again between E9 and E10, *ODist* is varied, which doesn't affect the relation between *Alpha* and *Theta2*.

E11: This experiment is crucial for the success of S10. Before he makes a prediction he looks back to E10 and notices that *Theta2/Theta1* = 2.5. He applies this relation to the current design which results in a correct prediction. Transformed in prediction format, this relation is represented as H10.

Table 4.12

All hypotheses of subject S10 expressed as rules

H1	IF <any>	
	THEN Ray is bent	
H2	IF <any>	
	THEN Ray goes slightly upwards	
H3	IF <any>	
	THEN ray goes very slightly upwards	
H4	IF (Substance Flint)	
	THEN Ray is bent more drastically towards the Normal than in case of (Substance Glass)	
H5	IF <any>	
	THEN Theta2 = 4 degrees (up)	
H6	IF (Alpha = positive)	
	THEN Theta2 = positive value	
H7	IF (Alpha = negative)	
	THEN Theta 2 = negative value	
H8	IF <any>	
	THEN as Alpha increases Theta2 increases	
H9	IF (Substance Glass) (Alpha 15)	
	THEN (Theta2=6)	
H10	IF <any>	
	THEN Theta2 = -Theta1/2.5[a]	
H11	IF <any>	
	THEN Theta1 = -Alpha	
H12	IF (Substance Flint)	
	THEN Theta2 = -Theta1/3.0	
H13	IF (Substance Glass)	
	THEN Theta2 = -Theta1/2.5	
H14	IF <any>	
	THEN As Theta2 increases Gamma increases	
H15	IF <any>	
	THEN Alpha has opposite sign of Theta1.	
H16	IF <any>	
	THEN the angle Theta2 increases from Glass to FlintGlass to Diamond	
H17	IF (Substance Diamond)	
	THEN Theta2 = -Theta1/3.5	
H18	IF (Substance Glass) (Surface Concave)	
	THEN Theta2 = Theta1/2.5 (the signs are changed)	

Note. If identical condition parts occur in different rules, the later rule overrides its predecessor.
[a] The minus sign has to do with a peculiarity of specifying angles in REFRACT: Theta2 had the different sign of Theta1.

E12: Is a kind of 'regression'. The learner decides to go with a graphical *Ray* prediction despite the fact that he could use H10 here effectively.

E13: Here S10 looks actively for other numerical relations between variables. He finds that *Theta1* has always a different sign than *Alpha*, and that *Gamma* is equal to *Theta2* in E13. However, both findings are correctly qualified as irrelevant, since they relate an independent to an independent and a dependent to a dependent variable, respectively.

E14: This experiment reveals that S10 really works in a discrimination mode, since H10 which was built in the context of an experiment with substance type *glass* is applied here where substance is *flintglass*: A typical overgeneralization that results in an incorrect prediction. As a result, H10 is probably discriminated leading to the specialization H13, and the phenomenon observed is encoded in form of H12.

E15-E16: The new hypothesis H12 is applied in two experiments, the first one, E15, being redundant again. In E16, the learner looks for additional numerical relations in the data.

E24: Another 'regression' to an unprecise prediction type, this time explainable inasmuch as this experiment was the start of Ss third learning session. That means, there was more than one day interspersed between E16 and E24. Also, the lens experiments done in the meantime may have had a distracting effect. Alternatively, S10 might have been uncertain whether surface form (which is for the first time *not plane*) has an effect on the relation between *Theta1* and *Theta2*. If so, this experiment shows him that this is not the case.

E25-E26: In E25 and E26, the knowledge from former experiments is available again: H12 and H13 are applied successfully.

E27-E28: This is the first experiment with substance *diamond*. S10 knows by now that substance type has an effect on the relation between *Theta1* and *Theta2* and selects a graphical prediction modus since he doesn't know about the constant for *diamond* yet. The relation is quickly found, resulting in the construction of H17. This hypothesis is correctly applied in E28.

E29-E31: The sign changes related to the change from convex to concave are found.

In short, S10 focuses mostly on quantitative aspects and restricts hypotheses' scope by means of discrimination learning. In order to find functional relations, he often uses the result of the comparison between predicted and correct value for *Theta2* to improve his predictions. For example, based on the relation captured in H10, *Theta2* = *Theta1*/2.5, the subject predicted in E14 a wrong value, realized that the correct relation there is *Theta2* = *Theta1*/3.0, and used this new relation in E25. Further, S10 does not generate more complex equations than necessary for prediction making. In particular, no index of refraction is induced. However, S10 does attempt at times to find additional numerical relations even in experiments where he made a correct prediction.

Before discussing the implications of S03's and S10's behavior in relation to building computational models, let me present the results of some further analyses of the two subjects' data.

4.4.2 Search Through The Experiment Space

Design Selection

Figures 4.3 and 4.4 show the path of S03 and S10, respectively, through the space of possible designs. The first obvious difference is found in the total number of experiments (S03 made 12, S10 24 experiments with single-surface refraction). To correct for this difference I will use only the first 12 experiments of S10 in many of the following comparisons.

The difference between the two with respect to the design selection strategy is obvious: S03 varied more variables at a time (average 3.6 variables changed between two experiments) than S10 (average: 1.6). S10 repeats designs in two cases with the argument that he *want's to get it right* before he continues. S10 produced similar[2] designs until he found a hypothesis that led to a precise prediction.

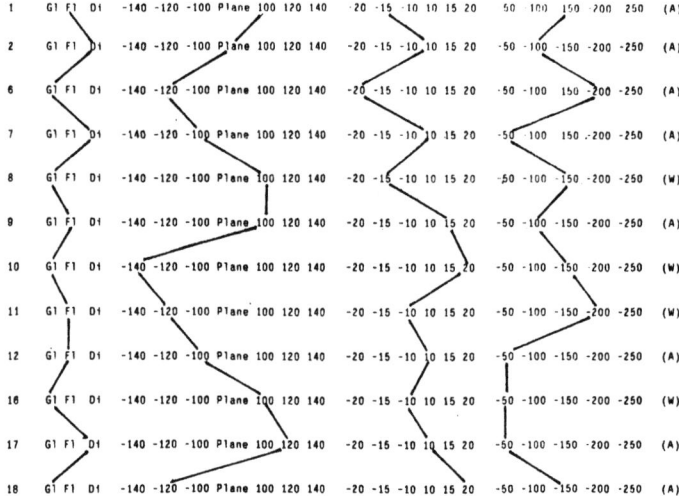

Fig. 4.3: Design sequence of Subject S03. From left to right: medium type, radius, *Alpha* and object distance.

---

[2] Two designs in REFRACT are similar if the substance is the same and the form of the surface falls in the same category, i.e., is either convex (positive radius), concave (negative radius), or plane.

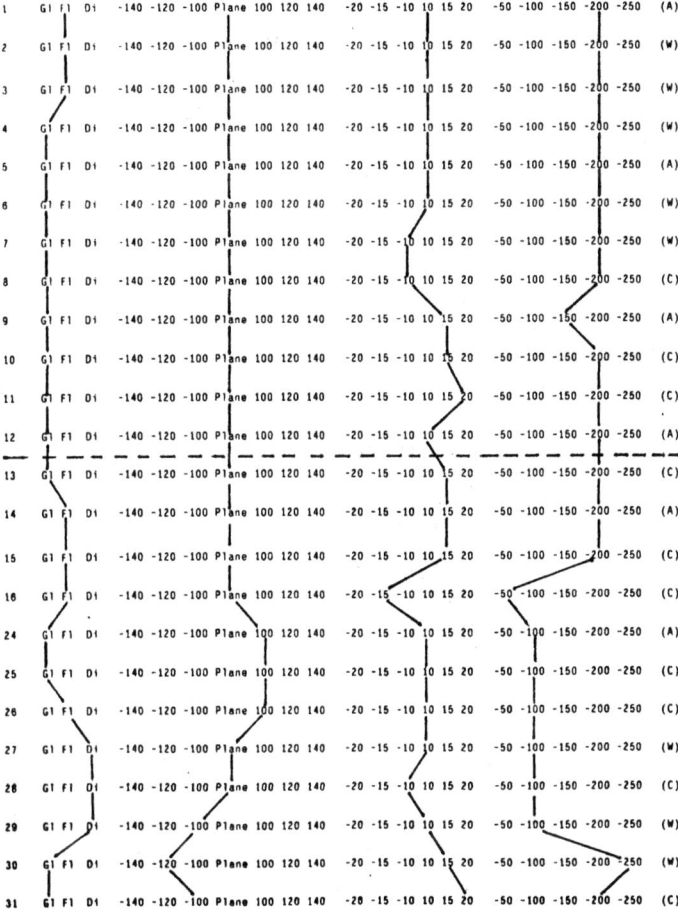

Fig.4.4: Design sequence of Subject S10. From left to right: medium type, radius, *Alpha* and object distance.

Design Reasons

From analyzing the reasons the two Ss stated for selecting a design[3], I found that S03 gives as many design related reasons (e.g., *I didn't try diamond yet*) as he gives hypothesis related reasons (*I want to test my assumption about the Normal*), while S10 utters mainly hypothesis related reasons (cf. Figure 4.5). Also, S03 selects more often design combinations without giving a specific reason than does S10. This indicates that S03 is more an 'experimenter' whereas S10 is more

---

[3] Ss were asked in every experiment explicitly for reasons after they had assembled their design.

after the design is assembled, whereas S10 tends to create experiments that relate to an existing hypothesis. His tendency to repeat designs might even be an indicator for something like a confirmation bias.

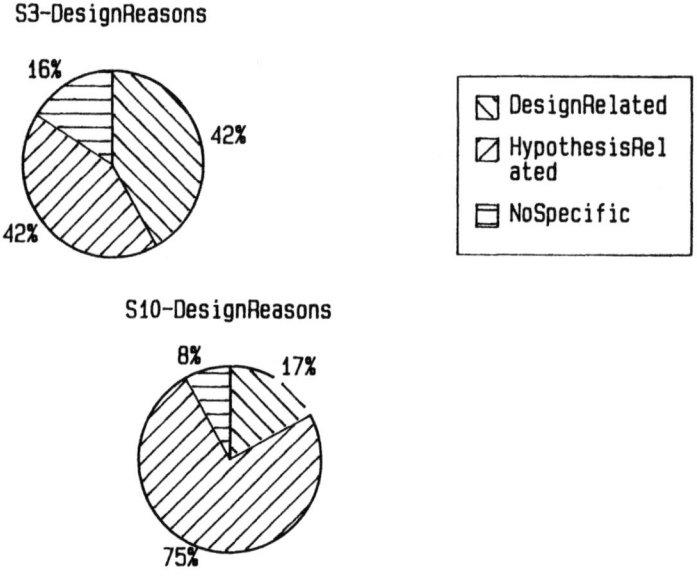

Fig 4.5: Verbal design reasons for S03 and S10

Another finding that bears more directly on this point results from the following analysis. In the verbal protocols, I counted how often a hypothesis was generated immediately *after* a design was created, because this is an indicator that the design drove the hypothesis generation process. For S03, I found a total of 8 such situations, or 67%, for S10 during his first 12 experiments only 3 (25%), showing that S03 is more data-driven than S10.

4.4.3 Hypothesis Modification

Prediction Type Used

In Figure 4.1, the prediction type used in the experiments was plotted against the experiment sequence. We saw that S03 uses basically *Area* and *Ray* predictions, i.e., graphical ones, and exhibits only a slight trend towards more precise predictions in the course of his learning sequence (by using more *Ray* predictions in later experiments, see Figure 4.1). In contrast, S10 starts out with the least precise prediction type (*Area*), but beginning with experiment number 6 he generates mainly numeric predictions, occasionally falling back on graphical *Ray* predictions.

As pointed out in Section 4.3.2, for S10 the appearance of *Ray* predictions is correlated with changes in design (see Figure 4.4): S10 utilizes the less precise prediction type when he encounters a 'new' design situation, i.e., when changes in substance and/or surface occur. In Figure 4.6, the percentages for the three prediction type categories are displayed.

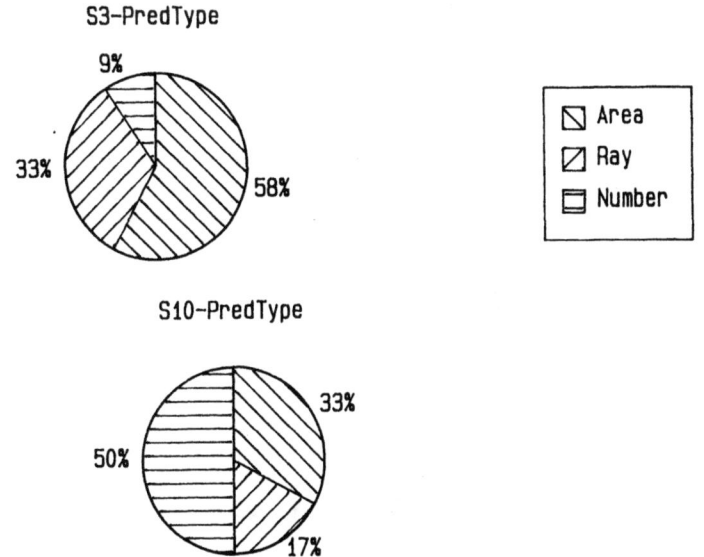

Fig. 4.6: Prediction type preferences for S03 and S10

Search for numerical features

Since predictions are based on hypotheses, the difference in prediction type preference between S03 and S10 cannot explain why S03 relates mainly to graphical features and S10 more to numbers. A possible explanation is that S10 might know better how to analyze numerical data and finds it therefore useful to consider them. An indication for this comes from an analysis shown in Figure 4.7, where the number of numeric relations tested is plotted against the experiment sequence. This highlights an important aspect of search through the hypothesis space: search for numerical relations. Clearly, S10 searches more in this direction (total: 10, average: .80 per experiment) than does S03 (total: 5, average .41).

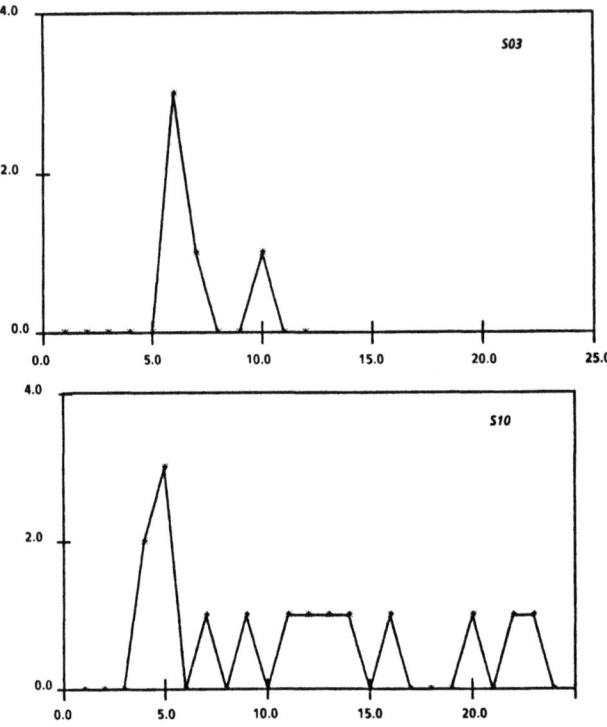

Fig. 4.7: Frequency of looking for numerical relations over experiments for S03 and S10

The results of the search through the hypothesis space are the hypotheses themselves, which have been listed in Tables 4.11 and 4.12. In the following, the hypotheses generated by the two subjects are compared with respect to three criteria: correctness, parsimony, and connectedness.

Correctness

Both subjects produced about the same number of hypotheses (S03: 15, S10: 18). However, these result from different numbers of experiments (S03: 12, S10: 24) and the contents of each S's hypotheses are vastly different. S10's final hypotheses are both correct and precise. Essentially, he ends up with the law of refraction as it is approximated in REFRACT (see hypotheses H12, H13 and H17 in Table 4.12). The only thing that's missing to make the generalization complete is assigning a variable for the index of refraction; the subject stops with hypotheses that have numerical constants for the index of refraction for each substance. That is, however, all that is needed to generate correct predictions. S03, on the other hand, does not have any hypothesis that links dependent to independent variables in a correct way and basically ends up having a set of hypotheses based on graphical features of the experiment.

Parsimony

Another aspect of the knowledge structure besides validity is the parsimony of the representation. I use the concept of a *default hierarchy* as described in Holland et al. (1987) to capture this feature. This means I try to identify the set of final rules a subject uses for prediction. Within this set, one rule may serve as the "default", while the others cover "exceptions" from the default rule. The hierarchies for S03 and S10 are displayed in Table 4.13

Table 4.13

Final hypotheses of S03 and S10 organized as default hierarchies

---

S03:

Default Rule: IF <any> THEN Ray is bent.

Rules for exceptions:

- (1) IF (Surface Plane) (Substance Glass) THEN Ray goes straight through.
- (2) IF (Surface Concave) THEN Ray is bent out slightly.
- (3) IF (Surface Concave) (Substance Diamond) THEN Ray is bent out and very close to the Normal.
- (4) IF (Surface Convex) THEN Ray is bent in.
- (5) IF (Surface Diamond) THEN Ray is bent in.
- (6) IF (Surface Concave) (ODist -150) THEN Ray is bent in and runs slightly under the Normal.
- (7) IF (Surface Concave) (more conditions) THEN Ray is bent in and runs slightly under the Normal.

S10:

Default Rule: IF <any> THEN Ray is bent.

Rules for exceptions:

- (1) IF (Substance Glass) THEN Theta2 = Theta1/2.5
- (2) IF (Substance Flint) THEN Theta2 = Theta1/3.0
- (3) IF (Substance Diamond) THEN Theta2 = Theta1/3.5

S10's final hypotheses are organized in form of one default rule and three exception rules. As said before, these rules cover the law of refraction as it is approximated in REFRACT. S03's final knowledge structure, on the other hand, includes more rules representing exceptional cases. This is so because he does not find the numerical relations that make an economic formulation of prediction rules possible.

The Tables 4.14 and 4.15 show, in a more condensed way and for all hypotheses, the number of conditions in a hypothesis and the degree of precision. The table contains in the columns a classification of phenomenon descriptions according to whether they are stated as a description of the path of the ray (i.e., a graphical description), as constraints on the value of a dependent variable, as a covariation statement between dependent and independent variables, or as an equation. Thus, this dimension covers the precision of a hypothesis, increasing from left to right. Along the rows, an estimate of the number of conditions appearing on hypotheses' left-hand side is given. Going from top to bottom, hypotheses become more and more specific. Both tables are based on Table 4.11 and 4.12, respectively. Clearly, S10's hypotheses are in the average both more general and more precise.

Table 4.14

Number of conditions against precision for S03's hypotheses

No. Cond	Phenomenon Type			
	Path	Constraint	Covariation	Equation
0	1,2	9, 10, 11, 14		
1	4,5,6, 8, 15			
2	3, 7, 12, 13			
>2				

Note. Numbers refer to the hypotheses listing in Table 4.11

Table 4.15

Number of conditions against precision for S10's hypotheses

	Phenomenon Type			
No. Cond	Path	Constraint	Covariation	Equation
0	1,2,3		8, 14, 15, 16	10, 11
1		5	4,6,7	12, 13, 17
2		9		18
>2				

Note. Numbers refer to the hypotheses listing in Table 4.12

Connectedness

A final aspect of the knowledge structure analyzed is the connectedness between the microworld variables as seen by the students. In Figure 4.8, a rough measure of connectedness is shown. It was generated by analyzing the hypotheses from Table 4.11 and 4.12 according to the relations noted between variables. If a relation was mentioned at least once, a line was drawn between the two variables. Not surprisingly, S10 has many more relations between the variables than does S03. This is, of course, also a function of the number of hypothesis stated.

4.4.4 Summary

The main lesson to be learned from this protocol analysis is that the two subjects attended to different aspects of information from the environment and that this had effects. After a few initial experiments, S10 searched actively for numerical information and utilized this to form hypotheses that finally took the form of equations, whereas S03 did often not go beyond the graphical information provided on the screen. Processing of information and final knowledge structure both reflect this difference. In particular, S03 needs more prediction rules to capture what appears to be exceptional cases, whereas S10 can represent the domain finally with a few precise rules based on numerical relations between the angle of incidence, substance type and the angle of refraction.

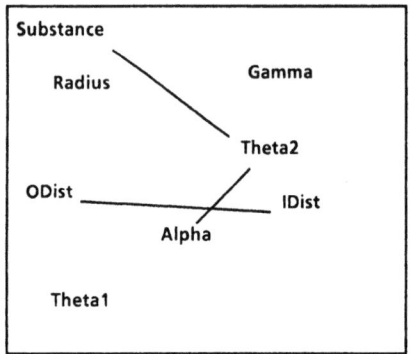
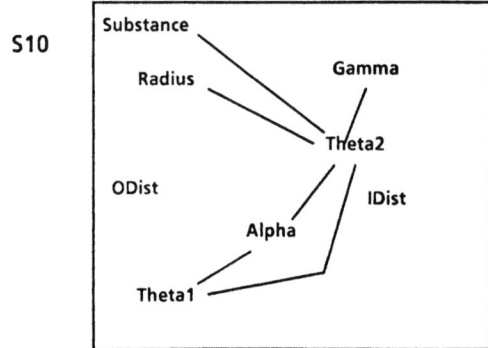

Fig. 4.8: Connections between variables for S03 and S10

I further found that the two subjects actually seem to employ a discrimination learning approach to hypotheses' scope modification. Moreover, they did induce only simple equations (if any). In order to describe the many differences between Ss economically, I will attempt to capture them (in Chapters 5 and 6) as being caused by differences with respect to three sources: problem representation, problem solving knowledge, and prior domain knowledge.

Problem Representation. The main source for differences I found concerns problem representation, i.e., the way subjects perceive the entities in the discovery world. As we have seen in analyzing two subjects' protocols and by looking for evidence in protocols of other subjects, a prominent difference was whether numeric information is included in their representations of experiments or not. Different subjects use different descriptors to encode what they perceive. The categories used by a learner to encode the information coming from the (microworld) environment determine the input for the inductive learning process. The nature of these observational statements has a significant influence on the quality of the learning process, both in terms of its efficiency and its outcome. Effects on the learning outcome result from the fact that the pictorial versus numerical representation are not informationally equivalent, i.e., it is not the case that "all of the information in the one (representation) is also inferable from the other" (Larkin & Simon, 1987). They have also different computational characteristics. For example, comparing two experiments in a graphical representation versus a numerical one will require different amounts of memory and processing time. In the next chapters, I will elaborate on such notions as 'knowledge structure', 'representation', and 'information processing'.

Problem Solving Knowledge in Rule Form. Problem solving knowledge comprises *domain-independent* knowledge. For example, knowledge about mathematical operators, how to build experimental designs, how to find trends in data, how to find functions in data, and heuristics such as *Improve your hypotheses before inquiring about new phenomena*. A second type of problem solving knowledge is *domain-specific*, such as knowledge about how to use the *NoteBook* tools.

Prior Domain-Specific Knowledge. This is the inevitable rest category. In particular, it refers to the correct or incorrect assumptions subjects have about refraction before they start working with REFRACT.

4.5 Conclusions

I reported observations on eight subjects who worked with REFRACT. This exploratory study was conducted in order to get insights into the specificities of discovery learning in REFRACT. I presented the observations in form of summative descriptions of subjects' final knowledge state, their prediction behavior, experiment construction skills, and features of their hypothesis generation processes. Only two subjects produced final hypotheses that capture the law of refraction as present in REFRACT quantitatively. The final knowledge state of the other discovery learners can either be described as being 'semi-quantitative' (hypotheses incorporate change information in the form of 'as x increases, y decreases') or 'qualitative-pictorial' ('Ray is bent upwards'). The lack of quantitative hypotheses in many subjects was traced back to shortcomings in designing experiments, lack of knowledge about how to analyze quantitative observations, and a preference to describe phenomena in 'pictorial' rather than numerical form.

Since the formulation of hypotheses in quantitative versus pictorial form had such drastic effects on subjects discovery behavior and the discovery outcomes, I undertook a in-depth analysis of two subjects' verbal protocols; one subject was a 'pictorial' discoverer, the other one worked mostly in a 'numerical' mode. I described differences between these subjects with respect to their final knowledge structure (content, correctness, parsimony, connectedness) and analyzed these differences in the light of variations in discovery behavior: different approaches to design selection, prediction behavior, and search for numerical features.

Finally, I explained the observed differences between subjects' discovery behavior in terms of three factors: problem representation, problem-solving knowledge, and domain-specific knowledge. The complex interactions between these components and the dynamically changing microworld can be analyzed effectively when they are actually implemented in form of running computer programs. In particular, I want to model the effects different knowledge about how to find regularities in data and different forms of problem representation have on the hypotheses generated and on the process aspects of hypothesis generation, such as working memory load. The groundwork for such computer models of scientific discovery learning in REFRACT will be laid in the next chapter, where I analyze the discovery problem in terms of information-processing and problem-solving requirements. Concrete implementations will be described in Chapter 6.

5. Discovery Learning as Problem Solving - A Task Analysis

5.1 Introduction

I analyze the discovery problem in REFRACT within the framework of problem solving theory. A conceptual model of the task is developed by characterizing the problem space for REFRACT and the characteristics of search processes therein. The analysis focuses on issues of hypothesis generation and modification; the question of experiment generation is not treated in detail. Starting point of my considerations is the GRI model, which is extended and modified into HDD, a conceptual model of discovery learning in REFRACT. HDD, the *Hypothesis Driven Discoverer*, forms the basis for the computational models to be described in Chapter 6. Before I start with introducing the conceptual model, let me clarify the relation between the information-processing analysis done in this chapter and the empirical observations reported in the last chapter.

5.2 Constraints Based on Empirical Observations

In this and the following chapter that deal with computer modeling, I do not attempt to model any of the subjects from the study. Such a simulation would have to include too many idiosyncrasies in subjects' behavior that are due to features of the specific REFRACT version used in the study. Rather, I want to model certain aspects of discovery learning in REFRACT that I think are of more general interest; for example, how regularities can be discovered based on little knowledge about experimenting and data analysis; or what the problems are with discovering invariants when the basic representation of objects is not quantitative. My intention is to model abstract, 'prototypical' discovery learners, where a prototype stands for a specific way of dealing with the task in REFRACT. This kind of analysis should help to "define the differences in demands that different methods of task performance place upon the subject" (Simon, 1975), which is the main function of a task analysis.

Instead of being used for simulations, observations enter this task analysis in form of constraints on the discovery learning processes. The most important constraints that I infer from the subjects' discovery behavior (see Chapter 4) are: (a) do not assume that experiments are designed systematically; (b) do not assume a lot of knowledge about data analysis methods; (c) don't expect that hypotheses are consistent with all the evidence and with respect to each other; (d) do not assume that significant causal knowledge about optics is brought to the task; and (e) do not assume that experiments are necessarily represented in a numeric format. These constraints, combined with general knowledge about inductive reasoning, will in various forms be used in the next two chapters, where I provide problem-solving models of the discovery problem, first conceptually, then computationally.

The kind of discovery system that will obey these constraints (or parts of them) can not look like BACON (Langley et al., 1987), which requires a factorial design sequence and has at its command sophisticated data analysis strategies. It can also not resemble AM (Lenat, 1982) or SDDS (Klahr & Dunbar, 1988), two discovery systems that bring to bear a substantial amount of background knowledge. Rather, on an abstract level, the observations made on untrained students suggest a model of discovery that builds on a basic generate-and-test scheme for hypothesis generation, where the hypothesis generator is constrained by heuristics that utilize features of experiments. I use Simon & Lea's GRI model of rule induction as as starting point for my considerations since it includes a clear notion of the generate-and-test strategy.

5.3 Extending the General Rule Inducer

A learner in REFRACT is confronted with the problem to find out which regularities govern the behavior of the objects in the environment. Starting from single observations, his task is to infer more general assertions. More specifically, the following features characterize the learning task in REFRACT.

- The learner is assumed to be a 'novice', to have no knowledge about refraction phenomena besides some rough starting hypotheses such as "Light rays bend when they go from air into another medium", or wrong knowledge.
- The only source of information for the learner are the outcomes of simulated experiments. These experiments have to be designed by the student himself.
- The environment contains both relevant and irrelevant information. It is up to the student to find out which variables convey the relevant information.
- The task is solved when the learner is able to predict the path of rays correctly for all possible combinations of light rays and media. The student does not have to explain why things happen the way they do; his only concern is with predictive power.
- Since the data are acquired in an incremental fashion by running one experiment at a time and since the learner has to make predictions based on incomplete information, he often will have to revise hypotheses based on new evidence.

Let me try to formulate the rule induction problem in REFRACT in light of the GRI model and of BACON. I propose to see a learner that works with REFRACT as searching in two problem spaces, one for experiments and one for hypotheses. Thus, the two central kinds of problems the learner encounters repeatedly may be paraphrased as *Which design should be assembled for the current experiment?*, and *Which hypothesis should be generated (activated, modified)?* Starting with this basic assumption, certain extensions of Simon and Lea's model will be necessary to accommodate the more complex task imposed by REFRACT.

Descriptive Generalization. First, in REFRACT the rules to be generated take the form of hypotheses about the path of rays. These hypotheses are more complex than the simple classification rules sufficient for concept formation tasks. In the terminology of Michalski (1983), the problem in REFRACT is one of *descriptive generalization*. The goal is "to determine a general description (a law, a theory) characterizing a collection of observations" (ibid., p. 118). Compare this to the problem of concept learning, where the learner has to produce classification rules of the form *If stimulus s has attributes a with values v, then s is an instance of the class*. In the domain of REFRACT, the properties to be specified take the form of functional relations between variables, such as *As the angle Theta2 increases, the image distance IDist decreases*. In cases of successful learning, these relations will be mathematical in nature, specifying the value of an independent variable as the result of combining dependent variables, for example *Theta2 = Alpha/2*. The distinction between dependent and independent variables also points to the fact that generalizations are *predictive* in that the value of the dependent variable can be predicted given the other values appearing in the equation.

Condition Finding. In cases where descriptive generalizations are based on observations which are all immediately available to the learning mechanism, the system usually has to derive a universally quantified generalization, a law. For example, the algorithm to find constancies and trends as implemented in the BACON.3 program (Langley, 1979a) presupposes that all relevant observations have been gathered (based on a complete factorial design) before induction starts, and that an equation has to be found describing all observations correctly. Therefore, the system will not come into a state were it has to backtrack because of hypotheses not being consistent with the data. In situations where the observations to be learned from come in incrementally--as is the case in REFRACT--generalizations often have to be augmented with conditions indicating the range of observations they hold for. When these conditions are subject to modification during learning, processes of descriptive generalization (modifying a rule's right-hand side, i.e., function induction) have to be combined with processes of condition induction (modifications of a rule's left-hand side). The learning task in REFRACT requires a combination of techniques for function induction and condition induction. Such techniques are described in Langley (1979a) and Falkenhainer & Michalski (1986).

We might speculate that in the case of learning in REFRACT human learners might acquire a final knowledge state consisting of three rules, each describing the relation between the angle of incidence and the angle of refraction for one of the three substances *glass, flintglass,* and *diamond,* instead of a single rule, where substance is 'variabilized in' by postulating a hypothetical entity 'optical density'. While inferring such 'hidden entities' is possible for BACON.3 and higher BACON programs that simulate the competence of professional scientists, untrained students might not be ready for such an inductive inference.

Attribute Selection. Further extensions of the GRI model are also necessary since in our case the learner has to find out what the relevant attributes for the domain are. In REFRACT, not all of the relevant attributes are necessarily salient for the student since they are not displayed automatically (for example, the angles of incidence, *Theta1*, and the angle of refraction, *Theta2*, are not) and many of the other attributes are irrelevant. Discrimination of relevance may be more difficult than in standard concept acquisition tasks since there are many attributes to choose from.

The attributes used by a learner to encode the information coming from the (microworld) environment determine the input for the inductive learning process. This is the case because the learning processes will, of course, operate on a representation of the objects and relations in the learning environment. We can think of these (mental) representations as being formulated by means of a symbol system, thus forming a kind of observation language. The nature of these observational statements has a significant influence on the quality of the learning process, both in terms of its efficiency and its outcome. For instance, I know from observations on students working with REFRACT that some of them behaved as if they did encode an incoming ray with angle *Alpha* set to 15 degrees as *running upwards*, for example, while others seemed to represent it as *angle Alpha has value 15 degrees*. If we assume that a learner starts with a set of initial descriptors (for example, those that are most salient), we can characterize different demands on the learning process according to the *relevance* of initial descriptors for the learning problem. Three cases can be distinguished (Michalski, 1983):

(a) *Complete relevance.* All the initial descriptors are relevant and sufficient to formulate the inductive generalization H that covers all observations. In this case, the learner has to combine the initial descriptors into an assertion H.

(b) *Partial relevance.* Only some of the descriptors are relevant for the generalization. The learner has to find the relevant ones and must combine them into an inductive assertion.

(c) *Indirect relevance.* None of the initial descriptors can be used directly for generalizing, but among the initial descriptors there are some that can be used to construct descriptors that are directly relevant.

A learner in REFRACT has to deal with the problem of partial relevance. Not all of the descriptors provided by the environment are relevant. For example, the angles *Beta* and *Gamma* do not allow to form reasonable generalizations. The learner has thus to find those descriptors which are relevant, or perform *selective* inductive learning.

Informational Feedback. In REFRACT, feedback is given by displaying the correct outcome; it thus conveys much more information than the right/wrong classification used in the concept formation paradigm. Processing feedback in this case may comprise several substeps: (1) analysis of the feedback information, in particular, comparison of prediction and feedback; (2) modifying existing hypotheses on the basis of new evidence; (3) steering the generation of new hypotheses. The richness of feedback information provides the learner with a chance to search heuristically for generalizations, i.e., to base generation of new hypothesis on an analysis of the experiments done already.

Experiment Construction. Finally, extensions of the GRI model are necessary since generating instances in REFRACT means to assemble experimental designs; this is a more demanding task than just selecting an element (like *Big Red Triangle*) from a fixed set of instances. In addition, we can assume that hypothesis generation and experiment construction will be interdependent. For example, the search for the next experimental design may be related to the current best hypothesis. In order to account for the interrelations between these components, the learner has to connect them by means of heuristic rules that guide the two search processes. As mentioned in describing the GRI model of Simon and Lea, this interaction provides the basis for heuristically searching the experiment space.

Summary. Five issues were raised in order to adapt Simon and Lea's analysis of rule induction to the discovery task in REFRACT: (a) equations have to be induced, not classification rules; (b) conditions have to be attached to these equations; (c) induction in REFRACT requires attribute selection; (d) feedback is multivalued and must be actively processed; (e) experiments have to be constructed.

A computational model of learning in REFRACT that covers the above mentioned complexities may have three main interacting components, which I label: GENERATE-HYPOTHESIS, RUN-EXPERIMENT, and EVALUATE-HYPOTHESIS. The interaction of the three main components is indicated in Figure 5.1 and can be described as follows:

- The output from GENERATE-HYPOTHESIS is a hypothesis, a rule in IF...THEN form.

- RUN-EXPERIMENT first generates a design, then produces a prediction by applying the hypothesis to the design, and finally runs the experiment and evaluates its result. The output of RUN-EXPERIMENT is a description of the discrepancies between the predicted ray path and the actual ray path.

- EVALUATE-HYPOTHESIS decides whether the current hypothesis should be pursued, modified, accepted or rejected.

This outline of the main components should be taken as a first hypothesis about the basic structure of my cognitive simulation model, the *Hypothesis-Driven Discoverer* (HDD). On this abstract level of decomposition many details are left open. For example, it is not specified whether a hypothesis has to be generated before the design step can be accomplished or whether hypotheses may not also be generated 'on demand'. Also, the model does not say how the information about experiments done up to a certain point is considered during experiment construction; and so on. Many of the more technical questions concerning the implementation of HDD are treated in Chapter 6. In this chapter, when I speak of 'HDD', of 'the system', or of 'the model' I refer in all cases to a conceptual information processing system, similar in spirit to Simon and Lea's description of GRI: The description is in terms of computational mechanisms, but on a level above of an actual computer implementation.

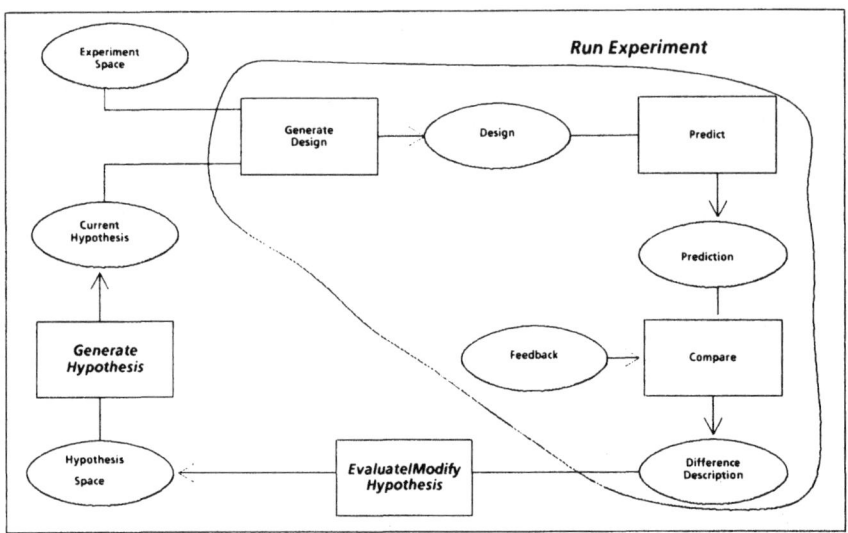

Fig. 5.1: The main conceptual components of HDD

A question not answered yet for my conceptual model of the task is: Given the basic components - generators for hypotheses and being processed? In other words, in which problem spaces should our artificial learner work? A problem space can be defined recursively as a set of primitive elements (objects, states) together with a set of operators that can generate new states by combining primitive elements. Thus, in defining a problem space for experiments and hypotheses, respectively, I have to specify the primitives and the operators that generate experimental designs and hypotheses.

5.4 Search in the Experiment Space

The notion of an experiment space, i. e., a mental representation for information from the learning environment, poses a difficult problem. Task analytical considerations depend necessarily on hypotheses about how subjects represent the external environment. A central aspect for system development is to define an 'attribute vocabulary' (VanLehn & Garlick, 1987) for the objects in the environment. In a complex task like the one encountered when working with REFRACT, we can not assume that all learners will encode the task relevant features in the same way. That leaves us with the question of how to characterize the various possible experiment spaces. An elegant solution is to come up with a general problem space description in the sense that all known representations humans use qualify as subspaces of this general description (Newell & Simon, 1972). However, this seems not to be a practical way to go in case of REFRACT; this environment is too complex and contains too much information as to allow for a simple and general description. And I know already that human discovery learners seem to employ at least two different forms of representing the experiment space: 'numerical' and 'pictorial' (cf. Chapter 4). For purposes of this task analysis, I will adapt a 'numerical' representation of experiments, in particular, one that is based on the *NoteBook* entries from REFRACT. This representation is more general and more easy to handle than any kind of 'qualitative' representation format.

States

A simple but effective representation of numerical data is one in terms of *feature vectors*, where objects are described using a fixed set of features or attributes that take on a finite set of values (Dietterich, 1982). For HDD, the basic format is an attribute-value pair notation, where each pair is indexed according to the context it is observed or generated in. Context can best be expressed in terms of experiment number. A data point can then be described as a triple:

(<exp#> <variable-name> <value>)

Variables from REFRACT are classified according to their *role*, *type*, and *status* in HDD. The *role* dimension distinguishes between independent, intervening and dependent variables. *Independent* variables (IVs) are those a subject can control directly. *Dependent* variables (DVs) can only be observed or measured and are assumed to depend on one or more of the IVs. A variable is called *intervening* if it influences dependent variables but can not be controlled directly.[1] See Table 5.1 for examples of the three roles. The *type* property specifies whether a variable takes a *numerical* or *nominal* form. In general, all observable variables in REFRACT with the exception of *substance* have numerical values. These values reflect either angles or the length of segments (distances). The *status* of a variable can either be *observable* or *theoretical*. A *theoretical* term is an attribute defined in terms of observable variables. For example, the expression *Theta1/ODist* is a theoretical term. Note that these terms can be made up of dependent or independent variables; as soon as a dependent variable is used in the definition of a theoretical term that term is treated as dependent. Thus, *Alpha/Beta* is an independent theoretical term, *Theta2/Alpha* a dependent theoretical term.

---

[1] The distinction between intervening and independent variables is not that important for the task analysis, since all the intervening variables are directly accessible in REFRACT. I will not further distinguish between the two.

Table 5.1

Classification of variables according to role and type

Variable	Role	Type
Alpha	Independent	Numeric
Beta	Intervening	Numeric
Gamma	Dependent	Numeric
IDist	Dependent	Numeric
ODist	Independent	Numeric
Radius	Independent	Numeric
Substance	Independent	Nominal
Theta1	Intervening	Numeric
Theta2	Dependent	Numeric

Operators

Two operations are required to gather data: *Design-Experiment* and *Get-Feedback*. Their realization is left open here since they are to be implemented as 'primitives', i.e., as simple fetch operations. The more interesting question is how search in the experiment space should be controlled.

Search Control

Many schemes to control the application of the design construction operator are possible. They vary according to the degree to which they take into account information about former experiments and existing hypotheses. Going from top to bottom in the following list, the experiment construction operators become more and more 'intelligent'. (a) Values for the independent variables could be randomly selected. A more clever system, however, would (b) take into account the experiment space, i.e., consider designs constructed already: values for independent variables would be randomly selected, but construction of an identical design would be avoided. This constitutes a generate-and-test strategy on side of experiment construction. Another possibility is to (c) vary IV values according to a fixed scheme, e.g., construction of a factorial design, or following the pattern of menus in REFRACT. At a most advanced level, (d) the hypothesis space would be taken into account, that is, search would be heuristically controlled. Design construction depends then on the current hypothesis (or hypotheses) and on knowledge about designs assembled already. A typical problem in this context is *ambiguity reduction*: Given two or more hypotheses with the same condition part but different phenomenon descriptions, design an experiment that can identify the correct hypothesis. Furthermore, decisions such as whether to look for confirmatory or discriminative evidence become relevant.

To get a feeling for the size of the problem space under very elementary conditions, let us assume a simple operator control scheme for searching the data space, one that generates a complete factorial design. The size of the resulting data space can then be calculated according to

$$S = V_1 * V_2 * \ldots * V_{n-1} * V_n,$$

where S is the number of value combinations in case of n independent variables with V_k values for each variable. For REFRACT, that results in

```
S = Substance levels * Surface level * Alpha levels * ODist levels
  = 3 * 6 * 6 * 5  = 540
```

combinations.

I will not further elaborate on these possibilities since design construction is not the central issue of my analysis (for recent work along this direction see Klahr & Dunbar, 1987; Kulkarni & Simon, 1988). In the implementations described in the next chapter, the problems related to design construction are actually avoided by assuming that experimental data are given to the programs as input. The following table summarizes the definitions in Backus-Nauer-Form (BNF).

Table 5.2

A formal characterization of the experiment space

State Description:	
<experiment>	<experiment><data-point>
<data-point>	(<exp-label><variable-label><variable-value>)
<exp-label>	exp-&Integer[a]
<variable-label>	Alpha\Beta[b]\Gamma\Theta1\Theta2\ODist, IDist\Radius\Substance
<variable-value>	Glass\FlintGlass\Diamond\plane\Numbers
Operators:	Get-Design, Get-Feedback

[a] The ampersand denotes character concatenation
[b] The backslash stands for 'or'.

5.5 Search in the Hypothesis Space

I assume that humans will learn in REFRACT by creating and modifying hypotheses. Correspondingly, generation and tuning of rules (as formalizations of hypotheses) are the central learning processes in HDD. *Generation* of hypotheses is based on existing hypotheses and information about experiments. The rule generation task can be subdivided into two subproblems: Induction of phenomenon descriptions and modifications of conditions. Phenomenon descriptions form the right-hand side of a hypothesis-rule, conditions controlling the application of a phenomenon description appear on the left side. Hypothesis *tuning* does not result in a new rule but in the change of a numerical *strength* parameter that is associated with each hypothesis-rule.

States

I distinguish between *hypotheses* and *predictions*; a hypothesis is a descriptive generalization over a set of observations; a prediction results from applying a hypothesis to a particular experiment with specific design constraints. A single hypothesis can thus account for several predictions. Syntactically, a hypothesis (about a refraction phenomenon) has two distinct parts: A set of *conditions* (which may be empty) specifying the scope of situations the hypothesis holds for, and a part that tells what will happen to a ray under these circumstances: the *phenomenon description*. Such a chunk of information can be conveniently expressed in rule form:

```
IF (attribute1 val) (attribute2 val) ... (attributek val)
THEN <phenomenon description>
```

Figure 5.2 illustrates the relation between hypotheses, predictions and production rules.

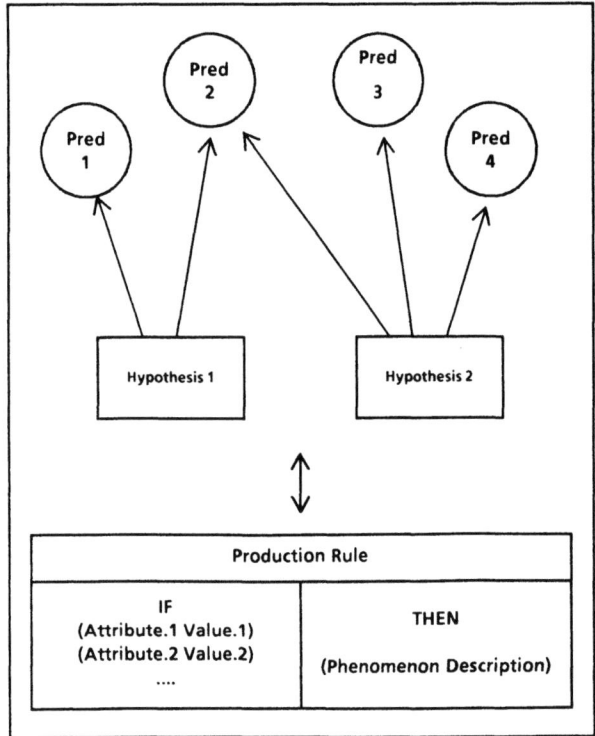

Fig. 5.2: Predictions, hypotheses and production rules

The condition part is represented as a sequence of independent variables and their values. The scope of a hypothesis can thus be more or less restricted depending on how many IV values are specified. Differences in scope can be expressed by means of using pattern variables as the value of independent variables. For instance, hypothesis (1) is more general than hypothesis (2):

(1)   IF (Substance =v1) (Alpha 10) THEN <...>
(2)   IF (Substance Glass) (Alpha 10) THEN <...>

The term '=v1' is a place holder indicating that any value for the attribute *Substance* can appear in this position. In this representation the learning system modifies the 'fillers' of the attribute 'slots' by replacing constants with variables (generalization) or vice versa (discrimination). The attributes themselves are not changed. Alternatively, or in addition, one might perform generalization/discrimination by removing/adding attribute-value pairs from a condition. This second meaning of generalization/discrimination is not used here. Note that the variable-replacement technique implies that a system using this representation 'knows' in some sense about the independent variables that can be part of a hypothesis' condition. For instance, it knows that there is an independent variable *Alpha*, but that its value does not matter. This kind of knowledge can be assumed for a learner in REFRACT, since he has to design experiments by explicitly selecting values for all the independent variables.[2]

A second implication of this representation format is that hypothesis-rules can be directly matched against experiment descriptions. Hypotheses do not have to be transformed in some way in order to apply to experiments.

After having specified what counts as a state in the hypothesis space, we have to define operators that can generate these states. Since states are production rules with a scope description as their left-hand side and a phenomenon description as their right-hand side, operators to create scope and phenomenon descriptions need to be developed. Before I turn to these operators, a general word on the search in the hypothesis space is in place.

5.5.1 Dimensions of Search

Langley (no year) discusses a number of dimensions along which different learning techniques vary in the context of concept learning: model-driven vs. data-driven, search direction, search control. I will adapt his analysis to the more general case of scientific discovery as instantiated by the learning task in REFRACT.

Model-Driven vs. Data-driven Search. A first and very basic distinction is whether the search in

---

[2] An exception is the intervening variable *Theta1*, which can not be directly controlled by the learner; however, in order to keep the representation consistent and since Theta1 can easily be accessed by the discovery system I do not distinguish it from the independent variables.

the hypothesis space is model-driven or data-driven.

> In the *model-driven* approach, one employs some knowledge about the content area to generate plausible hypotheses, and then uses the available data to evaluate these hypotheses. In the *data-driven* approach, one employs the data itself to generate acceptable hypotheses, usually in such a way as to guarantee that these hypotheses accurately summarize the data. In other words, the former method uses data only in the *test* stage of the search process, while the later uses data in the *generate* stage" (ibid., p.2).

In general, I assume that learning in REFRACT as simulated by HDD is more data-driven then model-driven. In other words, a learner in the discovery world who does not have any domain knowledge will most probably use the experimental data to base his hypotheses on, as did the subjects from my study. However, since the distinction is based on the amount of domain knowledge the learning system holds in addition to the information about experiments, this distinction is more a matter of degree than a real alternative.

HDD uses mainly data-driven operators to construct and modify hypothesis. There are operators to create phenomenon descriptions (for instance, trend detectors and function inducers), and there is an operator to modify the condition part of a hypothesis by discrimination. We can thus distinguish between two search processes within the hypothesis space: Search for phenomenon descriptions and search for conditions. The later search problem is much better analyzed in artificial intelligence research than the former, and the following distinctions apply primarily to the condition-finding problem.

Search Direction. In choosing the initial state for the condition finding problem, one has to decide whether the search should start from the most specific hypothesis or from the most general hypothesis. In the first case, condition finding operators remove conditions from the hypothesis, i.e., *generalize*; in the second case, they add conditions and hence perform *discrimination* learning. An approach that combines the two strategies was suggested by Mitchell (1977) and is called *version space* technique.

Search Control. At each point in the search tree, discrimination or generalization operators may produce more than one next state. For example, during generalization learning more than one hypothesis might be consistent with the data and a decision has to be made which one(s) to pursue. In the context of a learning system one must decide for a strategy to control application of the condition modifying operators. Two solutions are perceivable: One can select one of the candidate hypotheses and extend the part of the search tree below it: *depth-first* search. Alternatively, one can select all the candidates and expand them in parallel, thus performing a *breadth-first* search. A hypothesis is not further expanded if it leads to an error of commission (applying in cases where it should not be applied) in the case of generalization learning, or an error of omission (not applying in cases where it should be applied) when discrimination learning is performed. In case of depth-first search, the system will then backtrack to the closest parent node.

Another decision relevant for search control is whether search is to be performed exhaustively or heuristically guided. During *exhaustive* search control, both generalization and discrimination learners 'give up' on a hypothesis only when it led to an error. Under *heuristic* guidance, additional criteria are used to select a hypothesis (or a number of hypotheses when a breadth-first scheme is employed). For example, in the context of a breadth-first approach one might filter the candidate hypotheses by means of an *evaluation function* and expand only the hypotheses with a value above a certain threshold. This will lead to a *beam search*. I want to expand Langley's terminology by further distinguishing between local and global evaluation functions. A *local* evaluation function looks at a hypothesis as such, independent of its history or of other hypotheses. For instance, a value might be calculated for the complexity of the action side of a hypothesis rule. A *global* evaluation is one that takes into account the history of a hypothesis, for example, the number of times it was applied successfully, or its relation to instances, for example, a measure for how consistent a hypothesis is, given a number of instances.

Given this spectrum of possibilities, we have now to make decisions for HDD concerning the kind of operators employed, the search direction, and the search control. In the next section, these points are discussed for the equation induction problem. In section 5.5.3, the decisions for modifying hypotheses' scope are outlined.

5.5.2 Generating Phenomenon Descriptions

To induce the phenomenon part of a hypothesis for geometrical optics, the learner has to find a mapping from the set of input rays to the set of output rays. Such a mapping can, for example, be qualitative (*refracted ray is bent downwards*), semi-quantitative (*as the values for Alpha increase, the values for Theta2 decrease*), or quantitative (*IDist = ODist/.3*). In this task analysis, I will restrict myself to the second and third case, where phenomena descriptions take on a numerical form. In this search process, operators are applied that mathematically combine variables representing terms to form new terms. The construction of qualitative, 'pictorial' hypotheses as they were also uttered by subjects working with REFRACT (cf. Chapter 4) is much harder to handle computationally, as can be seen from the research on qualitative physics (e.g., Hobbs & Moore, 1985). I will make some suggestions regarding qualitative hypotheses at the end of Chapter 6.

Previous Research on Function Induction

Besides the already mentioned programs BACON and ABACUS, two earlier systems are relevant. Huesman & Cheng (1973) studied a mathematical induction task where they used two variables, five connectives (addition, subtraction, multiplication, division, and exponentiation), and constants from 1 to 4. In each problem, subjects were presented with six data points: pairs of values for the independent variable and the function's value. They had to determine a mathematical function that could have generated the data points. Using observations on subjects' reaction time and verbal reports as well as computer simulation techniques, Huesman and Cheng found that "induction is a heuristically directed generate-and-test process. The order in which hypotheses are generated is mostly independent of the data. Each hypothesis is maintained until it is negated, but a false hypothesis that matches part of the data is frequently retested" (p. 126). The basic heuristics used by their subjects for searching the space of functions were: (a) Try simpler functions first (here 'simple' means: those with fewer connectives); (b) consider addition, subtraction and multiplication before division and exponentiation; (c) for a given connective, consider constants before a variable. A similar generate-and-test method was proposed by Gerwin (1974) with supporting empirical data.

Huesman & Cheng (1973) and others (see Gerwin & Newsted (1977) for an overview) showed that a simple generate-and-test strategy is sufficient to solve elementary function induction tasks. Seen as a starting point to describe equation induction in REFRACT, their research has two important limitations. For one, subjects are presented with the data (numbers), they do not produce them on their own. Further, the functions describing the data were simple enough so that a generate-and-test strategy can work reasonably well. In might therefore be interesting to look what happens if subjects can select their own instances (Langley, 1979b). How to extend the simple generate-and-test scheme to deal with more complex functions has been discussed in the context of BACON (Langley et al., 1987) and ABACUS in Chapter 2.

Langley (1979b) studied a task similar to the one used in Huesman & Cheng (1973) with an important difference: Subjects were now allowed to select values for the independent variable on their own instead of being presented with a fixed set of value pairs. Langley's analysis, which builds conceptually on Simon & Lea's (1974) theory of rule induction and utilizes the production system technology, revealed that the task might more effectively be solved by not relying on a simple generate-and-test but using an abstract representation of the task. However, no empirical support was provided for this claim. Even though, this research provides us with an insightful task analysis of a prototypical function induction problem.

Operators for Inducing Equations

An important characteristic of function induction operators concerns the nature and the structure of the data they require as input. *Constructive* operators can transform input data before attempting to summarize them in an equation; *non-constructive* operators assume that the input data have the correct format already. An example for a constructive operator is an interval-construction operator that transforms the number sequence *1,2,3,4,5,6,7* into two intervals, *Low=(1,2,3)* and *High=(4,5,6,7)*. More important is whether operators work based on a kind of *closed-world* assumption; under this condition, it is assumed that all the data are known before the operator is applied. The operators have to account for all data points available and no backtracking will occur. When data come in incrementally, the closed-world assumption has to be relaxed and the system must be prepared to deal with wrong inductions (wrong in the sense that existing generalizations are contradicted by new evidence). A final feature of the data input is whether the function induction operators require a *fixed* input sequence (e.g., a factorial design) or not.

For equation induction in HDD, I use non-constructive operators which take numerical data as input and return an equation relating a dependent variable (*Theta2* or *IDist*) to one or more independent variables (*Alpha, ODist, Theta1*). Furthermore, the operators do not require that data come in in a specific sequence. Finally, and most important, the equation induction operators generate equations based on very little evidence making no closed-world assumption. This is important since in REFRACT hypotheses have to be produced immediately; their construction can not be delayed until a sufficient number of experiments has been gathered (recall that predictions have to be delivered in each experiment).

The operators for inducing equations will now be described as production rules. It is important to keep in mind that these rules are different from those mentioned so far, where with rules I always referred to hypotheses which are encoded in rule form. The rules that realize the function induction task are operators which generate hypothesis-rules. If the context is clear, I will in both cases continue to speak of 'rules'. Otherwise, the rules that encode hypotheses will be explicitly coined 'hypothesis-rules'. Another terminological distinction is in place here. In this chapter, I will describe some of the problem-solving and learning *operators* HDD has available in a verbal *rule-*format, i.e., as *If...Then...* structures. In the next chapter, these rule descriptions will be translated into a specific production system language and will then become *production rules*.

The rules responsible for function induction take as input information about two experiments in sequence (the current and the last one) and return equations describing the mathematical relationships between an independent variable (like the *Angle of Incidence*) and an dependent variable (like the *Angle of Refraction*). A two step algorithm is employed that first discovers trends and then equations (see Figure 5.3).

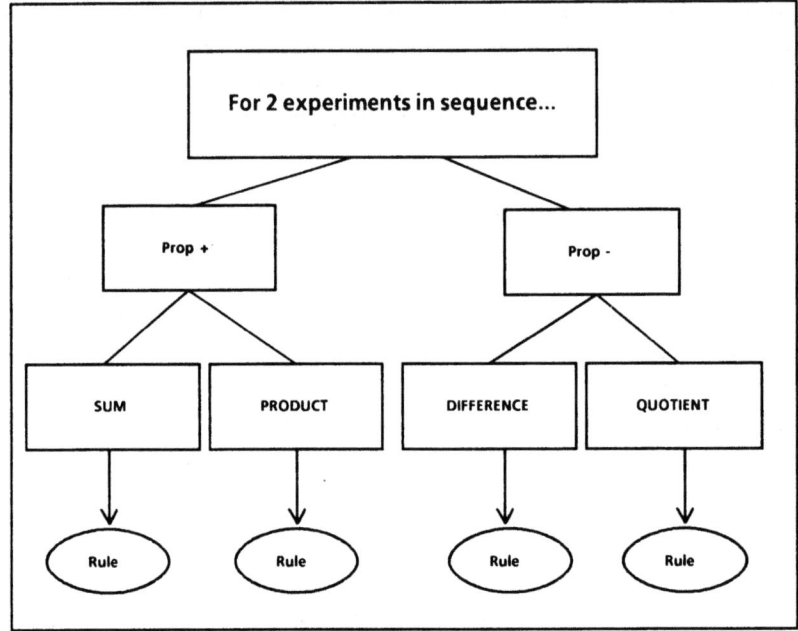

Fig. 5.3: Outline of the function induction step

Trends are discovered by means of four rules:

Trend.Direct1
```
IF     The goal is to find trends,
       and =d is a dependent variable,
       and =i is an independent variable,
       and both are numeric,
       and both are of the same type,
       and the value of =d increases between the last and the current
       experiment
       and the value of =i increases between the last and the current
       experiment
THEN   assert that =d and =i are directly proportional related,
```

and a similar rule, *Trend.Direct2*, which discovers the case where an DV and an IV decrease together. In the same manner, indirect proportionality is discovered:

Trend.Inverse1
```
IF     the goal is to find trends,
       and =d is a dependent variable,
       and =i is an independent variable,
       and both are numeric,
       and both are of the same type,
       and the value of =d increases between the last and the current
       experiment
       and the value of =i decreases between the last and the current
       experiment
THEN   assert that =d and =i are inversely proportional related,
```

and a similar rule firing in the case where the DV decreases and the IV increases.

When information about trends has been found, additional rules can have their condition side satisfied: rules that construct equations. The following four productions are proposed.

Function.Sum
```
IF      the goal is to find functions,
        and there is an inversely proportional relation between dependent
        variable =d and independent variable =i
THEN    build a new hypothesis rule which states that =d can be predicted
        by adding a constant value c to =i where c amounts to the
        difference between the current values of =d and =i.
```

Function.Difference
```
IF      the goal is to find functions,
        and there is a directly proportional relation between dependent
        variable =d and independent variable =i
THEN    build a new hypothesis rule which states that =d can be predicted
        by subtracting a constant value c from =i where c amounts to the
        difference between the current values of =i and =d.
```

Function.Product
```
IF      the goal is to find functions,
        and there is an inversely proportional relation between dependent
        variable =d and independent variable =i
THEN    build a new hypothesis rule which states that =d can be predicted
        by multiplying the value of =d with a constant value c where c
        amounts to the quotient of the current values of =d and =i.
```

Function.Quotient
```
IF      the goal is to find functions,
        and there is a directly proportional relation between dependent
        variable =d and independent variable =i
THEN    build a new hypothesis rule which states that =d can be predicted
        by dividing =d with a constant value c where c amounts to the
        quotient of the current values of =i and =d.
```

A hypothesis rule generated by these operators may look like this:

```
IF      (Substance =any)
        (Alpha =any)
        ...
THEN    (Predict: Theta2 = Alpha/2.0)
```

The newly constructed hypothesis rules have a *maximal general* condition side. They are universally quantified statements, unconditional predictions, since their scope is not restricted to specific design patterns. In other words, it is assumed that the relation described in the phenomenon part holds under all circumstances. Such rules will often be wrong because they constitute overgeneralizations. The relation captured in the phenomenon part may only be true for certain experimental situations, for instance, only for *plane* medium surfaces. The discrimination learning procedure described in Section 5.5.3 is used to correct for overgeneralizations.

Search Control

Controlling the search for equations is problematic because it is difficult to derive good evaluation functions that indicate which equations are promising generalizations (Falkenhainer & Michalski, 1986). At the same time, the combinatorical problems of this search are paramount: Combining terms by mathematical operations in an unrestricted manner will lead to many terms that do not describe the data adequately. This problem is even intensified in our situation where the data the equation generation is based on may not tell the whole truth; data collection is not completed when equations have to be generated. For these reasons, application of equation induction operators must be tightly constrained. I use the following two constraints:

1. *Proportionality Heuristic*: If two variables exhibit a monotonically increasing/decreasing relationship, build a new variable as the product/quotient of the two monotonically related variables. In the strict sense, "variable x monotonically increases with y if the values of x always rise when the values of y rise while holding all other variables constant" (Falkenhainer & Michalski, 1986, 574). The ceteris paribus requirement is too strong for our purposes and is dropped.

2. *Units Compatibility Heuristic*: In the natural sciences, for two variables to be added or subtracted they must be of the same type, i.e., have the same physical units. This requirement is generalized here into a preference rule by assuming that a learning system in a physical domain such as optics will prefer to combine variables of the same type even where it would be possible to combine two of different type. I thus assume that the division and multiplication operators will be applied to variables of the same type before they are applied to variables of different type. For example, <angle>-<operator>-<angle> is considered before <angle>-<operator>-<distance>.

A further possibility would be to use an operator preference heuristic. Empirical findings (Huesman & Cheng, 1973; Langley, 1979b) suggest that human subjects search the space of numerical relations for simple function induction tasks from 'left to right', scanning through a list of numerical operators in the sequence: addition, subtraction, multiplication, division. Following such a sequence requires a certain kind of backtracking: If the first operator does not result in a successful hypothesis, apply the next one in the sequence. Such a preference heuristic restricts search in a model-driven way, whereas heuristics (1) and (2) express a data-driven bias: In order to generate hypotheses about functional relations, utilize information in the data available. For the moment, I want to avoid model-driven constraints as much as possible and see how far I can get with just the two heuristics from above. Also, I could not identify any clear preference heuristic in the data of subjects gathered in the context of the study from Chapter 4. For these reasons, no preference heuristic is employed during this task analysis.

The eight operators (four for trend detection, four for equation generation) guided by a heuristic search control scheme generate quite facile equations of the general form:

    <DV> = <IV> <Operator> <Constant>.

Other operators would be needed to construct more complex terms, for example, equations where two or more IVs are mathematically combined. However, the assumption is that equations are constructed by humans in a manner from the simple to the complex and that humans will be satisfied with a simple solution (Huesman & Cheng, 1973; Langley, 1979b; Simon, 1981). Equations of the general form described above allow to summarize experiments from REFRACT in a way so that all data points are described by three equations of the form:

```
Theta2 = Theta1 {TIMES or DIVIDED_BY} <Constant>,
```

with *Constant* taking a different value for substance *glass*, *flintglass*, and *diamond*. As we have seen in the study described in Chapter 4, learners in REFRACT do not go beyond this stage of summarizing data, do not generate a single equation that includes a hypothetical variable representing optical density, what would be Snell's law.

The heuristically guided generate-and-test search scheme expresses my impression that the human learners in the study did not process data extensively in order to find regularities, and did not necessarily attempt to keep old hypotheses consistent with newly acquired data. This assures that the computational requirements for generating equations are not overwhelming. Further, it is in agreement with Simon's (1981) claim that humans tend to "satisficing", i.e., to look for satisfactory solutions instead of optimal ones.

In Table 5.3, the representation format for the problem space of equations is summarized in a BNF grammar.

Table 5.3

A formal characterization of the equation space

State Description:	
<trend>	(<trend type> <IV> <DV>)
<trend type>	Prop$^+$\Prop$^-$
<equation>	<DV> = <IV> <operator> <c>
<DV>	Theta2\Gamma\IDist
<IV>	Alpha\Theta1\ODist\Radius
<operator>	+\-*\/
<c>	a Real
Operators:	Trend.Direct, Trend.Inverse, Function.Plus, Function.Difference, Function.Times, Function.Quotient

5.5.3 Modifying Hypothesis Scope

Modification of a hypothesis' scope is triggered by prediction faults, that is, it is a kind of *failure driven* learning[3]. The problem is to modify the component responsible for the prediction failure.

---

[3] Note that the case where the system is not able to deliver a prediction is also treated as a failure.

We thus assume that the three other problems self-modifying systems have to deal with: generating behavior, determining good and bad behavior, and assigning credit and blame, are solved. In other words, we assume that the system has produced data to learn from (here: experiments), it knows whether its behavior was successful or not (received feedback), and it knows which component of itself is responsible for the faulty behavior (identified the hypothesis behind the current prediction). How this is done in detail will be explained in Chapter 6.

Operators

Two operators are required, one for discrimination and one for strength tuning.

Discrimination. The condition side of a hypothesis rule can be modified by generalization and discrimination operators (Anderson et al., 1979; Michalski, 1983; Langley, 1987). I assume that a learning system working in REFRACT will primarily learn by *discrimination* learning, i.e., that it will start from general hypotheses and refine them under the influence of feedback. This assumption is based on the nature of the task in REFRACT: Learners have to produce predictions for *every* experiment they run; presumably, they will rely on knowledge acquired during the last experiments to derive a new prediction. This constitutes a bias towards formulating initially general hypotheses, a bias imposed on the learner by the task environment. It is further based on evidence gathered in my exploratory study, where hypothesis acquisition could be described as following a path from the general to the specific.

The triggering condition for discrimination is a wrong prediction. The hypothesis that led to the wrong prediction is identified and its scope becomes restricted. Specialization takes the form of *condition discrimination*. For example, let us assume that the following hypothesis was created in an experiment where the medium's substance was *diamond*:

```
H1:     IF (Substance =any)
        (Radius =any)
        (ODist =any)
        (Alpha =any)
        (Theta1 =any)
        THEN (Theta2 equals Theta1/3.0)
```

This hypothesis is overly general and will lead to prediction errors, but it is true for the experiment where it's generated in. Now suppose that the student in a later experiment makes the prediction *(Theta2 equals 3.3)* based on *H1* and makes the following observation:

O1: ... (Substance glass) (Theta2 5.0)) ...

He has to realize that its prediction is wrong, hence that the hypothesis is wrong and that it lacks one or more conditions. Such an error evokes the discrimination process to build more restricted ("conservative") versions of the original rule. To accomplish this, the learning procedure retrieves information about the last "good" context, that is, an experiment where *H1* was applied successfully (which may be the experiment it was created in), and compares it to the current "bad" situation, the experiment where a wrong prediction was made (cf. Figure 5.4). The goal of this comparison is to find differences between the "good" and the "bad" context. In this case, one such difference is found: in the good context, the variable *Substance* was bound to *diamond*, whereas in the bad context it is bound to *glass*. Therefore, the system creates a variant of *H1* with a new condition:

```
H1:     IF  (Substance diamond)
            (Radius  =any)
            (ODist   =any)
            (Alpha   =any)
            (Theta1  =any)
        THEN (Theta2 equals Theta1/3.0)
```

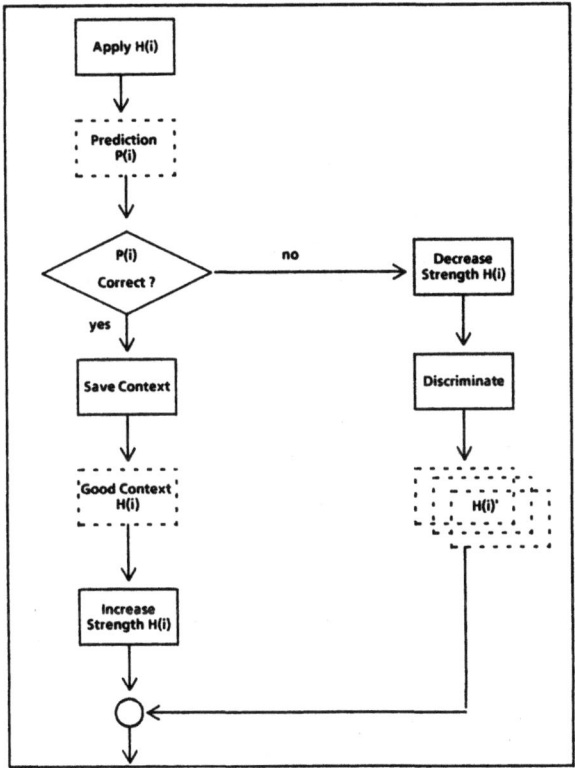

Fig. 5.4: Essentials of the discrimination step

This new, less general rule will not fire in situations where it was provenly wrong. Having derived a new rule that avoids the error made by its predecessor is not the whole story, however. The new rule has to compete with already existing rules, among them its more general predecessor (note that the faulty predecessor is not deleted but stays further in the system's knowledge base). Further, more than one new rule might be derived within a single discrimination step, dependent on how many differences between good and bad context were found. How can the system find out which of these candidate specializations is/are the most promising one(s)? What is needed is a mechanism that regulates competition among rules. Such a mechanism is provided in form of the *strength* parameter.

Strength Tuning. So far we considered changes of hypotheses (expressed as rules) that result in structural changes, that is, in new rules. A second learning operator does not change the form of a hypothesis, but the value of a quantitative attribute attached to it, the *strength* parameter. Strength reflects the relative success of a hypothesis in the following way: Whenever a hypothesis-rule is applied successfully, its strength parameter is increased. On the other hand, when the hypothesis leads to a wrong prediction (unsuccessful application) its strength is decreased. All hypothesis are augmented with an initial strength value when generated.

Structural modifications and strength modification are applied in combination. In discrimination learning, the unsuccessful application of a rule leads both to a change in its condition part (resulting in one or more specialized rules) and to a decrease of its strength parameter. In a production system, changing the conditions of rules will influence learning via the *match* process, while modifying quantitative attributes such as strength has effects on *conflict resolution*. In other words, while the condition part is considered by the production system interpreter to decide whether a hypotheses applies to the current situation, the strength parameter is used to decide between competing hypotheses.

Including strength revision in HDD as a learning process results in a system that is prepared to deal with incremental hypothesis refinement and noise. Its performance will not be 'brittle', but will improve or degrade gradually. Newly generated hypothesis rules have a low initial strength value and hence will not influence the systems performance right away. A hypothesis is considered for prediction derivation only when it holds for more than one experiment. This assures that hypotheses containing overly spurious correlations will not be considered for prediction making. Similarly, a hypothesis that has been applied successfully in a number of situations will not be completely abandoned when leading to a wrong prediction the first time, but be 'gracefully' degraded.

Search Control

Search for new conditions based on failure-triggered discrimination in combination with a strengthening scheme can be characterized as a *beam search* (Langley, 1987). Whenever the discrimination operator is called on a hypothesis rule, it will find all possible new conditions and add the resulting new rules to the hypothesis space. Not all of the possible rules are expanded by the discrimination operator but only the one(s) with the highest strength value. The strength filter functions as a heuristic criterion, saying that only the most successful rule(s) should be considered in the prediction step, and consequently only the most successful rule(s) will be further discriminated should they fail under more particular circumstances. The search is further constrained by giving newly generated rules a low initial strength value so that they do not compete with already established rules right away, but only if they have been rebuilt a couple of times (the exact parameters will be described in Chapter 6).

Discrimination-based condition learning has a number of advantages over the more often used generalization strategy: disjunctive rules can be acquired, the system can deal with noise in the data, and the system can back up to older rules should the environment change (see Langley, 1987). What makes this learning mechanism particularly interesting from my point of view, where discrimination learning is treated as a candidate learning mechanism for human discovery, is the fact that it requires only one piece of information to be stored with a hypothesis: the context of its last successful application[4]. But I think that the discrimination algorithm needs to be further constrained in order to account for limitations of human discrimination learning. I consider it not very plausible that rules are only weakened but never removed, i.e., never forgotten. (Note that giving up this assumption will limit the systems back-up capabilities.) Further, I reckon that humans will employ a very small 'branching factor' at each discriminated hypothesis node, that is, they will generate only a few, maybe only a single more specific hypothesis[5]. For example, the human discovery learners I studied did not seem to generate all possible variants of an incorrect hypothesis. An answer to the question whether this is due to the fact that they represent the application context in a highly selective way or consider only a few of the stored features for context comparison can only be speculative, given my current knowledge. Recently, Langley Gennari & Iba (1987) have suggested to describe human induction (in particular, concept formation) in terms of a much weaker search technique: as guided by a *hill-climbing* search control scheme where only a single node is expanded in each step, and where earlier parts of the search space are not retained in memory. However, for this strategy to work, the current state must be a fairly complex data structure, much more complex than single production rules.

The discrimination learning approach to rule induction satisfies a criterion Holland et al.

---

[4] A pointer to an external information structure holding that information--like the NoteBook--also suffices.

[5] This constraint was actually implemented in Anderson & Kline's (1979) version of ACT's discrimination learning mechanism.

(1986) consider crucial for any reasonable inductive system: gracefulness.

> We must emphasize the role of competition and the importance of graceful insertion of new rules into the cognitive system. Because of competition, the rules generated by the inductive mechanisms need not be universally correct, and the mechanisms themselves need not be very efficient. (...) Computationally overwhelming requirements for global consistency of rules and their predictions are avoided. Rules that do not improve performance are gradually eliminated by the refinement process. And all of this goes on without affecting system performance in well-practiced areas where strong rules dominate the competition. (p. 78)

Table 5.4 summarizes the assumptions concerning the form of conditions and their modifications in HDD as an BNF grammar.

Table 5.4

A formal characterization of the condition space

State Description:	
<scope>	(<condition.1><condition.2>...<condition.5>)
<condition>	(Exp <exp-label> <IV> <value>)
<exp-label>	exp-&Number
<IV>	Alpha\Theta1\ODist\Radius
<value>	<pattern variable>\<variable-value>
<pattern variable>	=iv1\=iv2\=iv3\=iv4\=iv5
<variable-value>	Glass, FlintGlass, Diamond, plane, Numbers
Operators:	Discriminate, Increase.Strength, Decrease.Strength

5.5.4 Summary

The problem of finding hypotheses was divided into two subproblems: creation of phenomenon descriptions and modification of scope descriptions. Since I assumed that phenomenon descriptions take the form of equations relating dependent and independent variables, the hypothesis generation problem was treated as the problem of inducing equations from experimental data, and as the problem of modifying the condition side of production rules based on feedback from the environment.

For the condition modification task, decisions concerning search direction, search control, and operators were discussed. I conceptualized the search for conditions on hypothesis-rules as being data-driven, progressing from general to more specific rules (discrimination learning), and being controlled by a best-first control regimen, with a measure for hypotheses' success as evaluation dimension: strength.

While I could build on a comprehensive body of research on condition learning both from artificial intelligence and psychological research, things are different with the generation of the action side of rules, i.e., descriptions of refraction phenomena in my case. Action induction is, by the very nature of the task, a much more domain-dependent process than condition modification. There is no way to tell in general how action sides should be created. In case of REFRACT, creating action descriptions amounts to inducing regularities from a set of experiments. In order to get started on the regularity finding problem, I decided to begin with as little assumptions as possible about what a system could know about the domain and about scientific discovery. Another simplifying assumption was that the description of regularities in REFRACT takes the form of equations, relating dependent to independent variables mathematically. Phenomenon induction - in the sense of equation induction - was then described as the application of simple trend-detection and function-creation operators to the representation of experimental data. The application of the operators is controlled by a heuristically guided generate-and-test scheme, where the heuristics capitalize on features in the experimental observations.

5.6 Conclusions

The discovery task in REFRACT was analyzed as problem solving in terms of search for experiments and hypotheses. I identified five major demands on the learner, human or artificial: (1) descriptive generalizations with predictive power have to be built; (2) since these generalizations are built on the basis of incomplete knowledge, they must often be restricted in their scope; therefore, the discovering system has to find conditions on overly general assertions; (3) relevant variables must be discriminated; (4) in order to improve its hypotheses, the (human and artificial) discovery system learning in REFRACT must process informational feedback; (5) experiments have to be constructed actively.

A conceptual model, HDD, was developed that treats discovery learning in REFRACT as a heuristic search through a space of experiments and through a space of candidate hypotheses which are generated by the application of various operators. The search through the experiment space was only touched briefly; my main focus lay on search for hypotheses. I proposed to represent hypotheses about refraction phenomena as production rules, where the left-hand side of the rule describes the scope of a hypothesis, the right-hand side describes a refraction phenomenon. The issue of finding conditions for hypothesis-rules is so important in HDD since hypotheses need to be generated before all the experiments are known to the learner and hypotheses will therefore often be overly general. Generation of rules on the basis of incomplete knowledge and revision of hypotheses that are no longer in accordance with the data are the general concerns of my analysis. For the phenomenon induction problem as well as the condition modification task the guiding idea was to trade ease of computation for completeness and consistency. Thus, it is not assured that a newly generated equations will be consistent with all the data available, nor does the discrimination algorithm take all observations into account.

Since they are sparkled over the chapter, a summary of the simplifying assumptions that underlie the conceptual model HDD is in place.

- Experimental observations are described in form of feature vectors. More complex representational schemes (such as a structural description of the domain entities) are not used.
- Experiments are given to the discovery system as input; issues of design construction are only touched briefly.
- The condition side of a hypothesis is represented as a feature vector. This means that the condition part of a hypothesis-rule can be matched directly against the description of an experiment. No complex transformations are required.
- The features that can appear in a hypothesis' condition side are known.
- Conditions for hypotheses are only searched one-directional: from the general to the specific (discrimination learning).
- The phenomenon side of a hypothesis takes the form of a simple equation relating a single dependent to a single independent variable.
- The search for equations is constrained by a heuristically controlled generate-and-test scheme.
- Two heuristics are employed: units compatibility and weak proportionality.

Building on these assumptions, HDD emerges as a discovery learner with the following main characteristics: (a) It represents the entities from the microworld quantitatively in form of feature vectors; (b) it searches for equations in a data-driven manner, taking into account an only small amount of information (precisely, information from two experiments); (c) it looks for regularities in the numerical data by identifying trends and by creating equations based thereon using four mathematical relations: *Plus, Difference, Times, Quotient*; it is satisfied with simple equations summarizing the data taken into account; in particular, it does not strive for equations that are consistent with all the experiments that were run; (d) it forms hypotheses about quantitative relations which are initially maximal general in scope; (e) it limits the scope of a hypothesis only when it led to a wrong prediction; (f) it restricts the scope of a hypothesis as little as possible, i.e., it attaches one condition before attaching two conditions. The result of the task analysis is summarized more formally in Table 5.5 that compiles the BNF notation for the experiment space and the hypothesis space.

It is obvious that this model of a discovery learner does not resemble any particular subject from the empirical study. Rather, it is a prototype of a learner that represents observations made in REFRACT quantitatively and looks for simple numerical regularities. It satisfies most of the initially mentioned constraints in that it does not rely on a particular sequence of experiments, employs only simple data analysis methods, does not rely on domain-specific or more general knowledge about physics, and in that is does not keep its hypotheses consistent (for example, hypothesis-rules with the same scope description may make different phenomenon predictions, and vice versa).

In the next chapter, I will describe a group of computer programs that implement the main aspects of the conceptual HDD model in various forms: phenomenon induction and condition modification, together comprising the hypothesis generation/modification aspects of discovery in REFRACT.

Table 5.5

A formal characterization of the problem space for discovery in REFRACT

EXPERIMENT SPACE
  State Description:

<experiment>	<experiment><data-point>
<data-point>	<exp-label><variable-label><variable-value>
<exp-label>	Exp-&Integer
<variable-label>	Alpha\Beta\Gamma\Theta1\Theta2\ODist, IDist\Radius\Substance
<variable-value>	Glass\FlintGlass\Diamond\plane\Numbers

Operators: Get-Design, Get-Feedback

HYPOTHESIS SPACE
  State Description:

<hypothesis>	((<H-Label> <scope> <equation>) <STRENGTH>)
<H-Label>	H&integer
<STRENGTH>	a Real
<scope>	(<condition.1><condition.2>...<condition.5>)
<condition>	(Exp <exp-label> <IV> <value>)
<exp-label>	exp-&Number
<value>	<pattern variable>\<variable-value>
<pattern variable>	=iv1\=iv2\=iv3\=iv4\=iv5
<trend>	(<trend type> <IV> <DV>
<trend type>	Prop$^+$\Prop$^-$
<equation>	<DV> = <IV> <operator> <c>
<DV>	Theta2\Gamma\IDist
<IV>	Alpha\Theta1\ODist\Radius
<operator>	+\-*\/
<c>	a Real

Operators: Trend.Direct, Trend.Inverse, Function.Plus, Function.Difference, Function.Times, Function.Quotient, Discriminate, Increase.Strength, Decrease.Strength

---

6. Computer Models of Hypothesis-Driven Discovery Learning

6.1 Introduction

In this chapter, I continue the task analysis from Chapter 5 by describing a series of computer models (in form of production systems) that solve the discovery problem in REFRACT, except for the experiment construction part. The notion of 'search for hypotheses' is made concrete by constructing programs that actually perform such a search, i.e., generate and test hypotheses. Further, I experiment with different approaches to deal with the discovery problem in REFRACT and describe the effects these approaches have on learning behavior and learning outcome. In particular, I compare a system that induces quantitative hypotheses in a fairly unconstrained manner with more constrained variants, and show how a system that generates qualitative hypotheses differs from the quantitative versions. These computer models correspond to idealized discovery learners, not to specific human subjects.

The computer models of discovery learning described on the following pages are all based on the same core architecture. Thus, these models can be seen as different versions of a single information-processing system, where each version corresponds to a different approach to deal with the discovery problem. Differences between the models are defined in terms of differences between the problem representation (numerical versus pictorial), differences with respect to function generation knowledge (more or less constrained operator application), and differences in specific architectural features. I begin with a computer model that is a fairly straightforward implementation of the conceptual HDD model developed in the last chapter: HDD-SH. Next, three more constrained versions of this discovery learning model are introduced (HDD-SHR, HDD-SHLR, HDD-SHOP), all working with the same (numerical) representation of the problem, but incorporating different assumptions about function induction. With HDD-SHOP, the notion of learning is extended to cover the acquisition of operator preferences. I close with discussing the implications of using a qualitative representation of the domain for discovery learning and introduce a simple production system that generates qualitative hypotheses.

6.2 A Quantitative Discovery Model: HDD-SH

HDD-SH is a fairly direct implementation of the conceptual discovery learner described in the task analysis in Chapter 5. In each experiment it derives a single prediction and tests a single hypothesis (therefore -*SH*). HDD-SH can be seen as a realization of the as of yet abstract notions that were summarized in Table 5.5. It is characterized as representing experiments numerically, generating new hypotheses based on the simple function-induction algorithm outlined in Section 5.5, and modifying hypotheses' scope by means of discrimination learning. I first describe the main components of HDD-SH and will then report the results of running HDD-SH with a sequence of experiments.

6.2.1 Model Description

A verbal description of a program is almost inevitably difficult to produce and to read. I will therefore stick in this description to the main points. For details, the reader is referred to Appendix VI.1 that contains the complete description of the architecture and the production system written in the PRISM production system language.

HDD-SH's performance is organized around a set of goals, corresponding roughly to the components of the SPFP sequence (Figure 6.1).

Step 1 - Designing an experiment. In the current implementation, the design step is left out. The program receives a design description as input instead of generating a design on its own.

Step 2 - Making a prediction. After having received a design description, the next goal is to make a prediction. Predictions are based on hypotheses. If more than one hypothesis can be applied given the current design description, a single one has to be selected.

Step 3 - Evaluating the prediction. The prediction is evaluated by comparing it with the actual ray path description as provided by the feedback step in REFRACT. The result is a description of the difference between predicted ray and actual ray or a statement that no difference was found. No distinction is made between *approximately correct* predictions and *wrong* predictions; both are treated as deviations from the correct case.

Step 4 - Evaluating and Modifying the Hypothesis. Based on the difference description, the hypothesis responsible for the current prediction is evaluated. A hypothesis can either be correct or wrong. When a hypothesis turns out wrong, a discrimination learning process is triggered that attempts to attach new conditions to the hypothesis building on information about the last correct and the current incorrect application of the hypothesis.

Step 5 - Generating new Hypothesis. If no hypothesis was available for prediction making[1] or in case that the current hypothesis resulted in a wrong prediction, the program attempts to create new hypotheses, i.e., rules incorporating a phenomenon description that has not been discovered so far. This is a two step process: first, trends are discovered; then functions are induced.

The goal sequence is realized by a set of production rules that run within a certain production system architecture. In order to define the production system architecture for HDD-SH, I specify the following components: interpreter, memories, learning mechanisms, representation language, declarative domain knowledge, and procedural knowledge.

---

[1] This is the case when the simulation starts and does not have any knowledge about refraction phenomena.

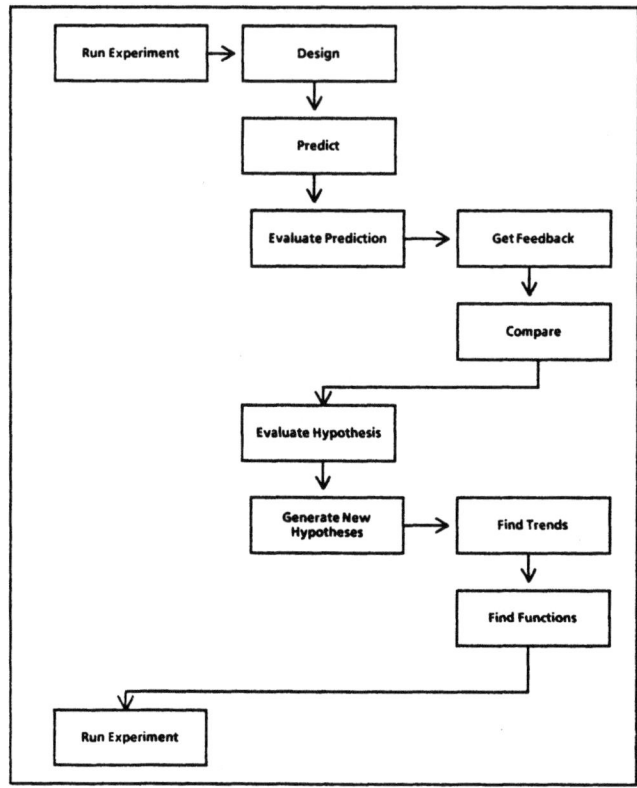

Fig. 6.1: The goal sequence realized by HDD-SH

The Interpreter

HDD-SH is written based on the production system shell PRISM, InterLisp version (Ohlsson & Langley, 1984; Opwis, Stumpf & Spada, 1987). PRISM is a very flexible interpreter that supports multiple memories, network and set structure for declarative memories, and rule learning, among other things.[2] In order to construct a specific production system within this shell, a number of decisions have to be made.

For declarative memories, the program designer has to decide how many memories of the declarative kind he wants and give them a name. Furthermore, he has to decide on the structure of declarative memories. Should a declarative memory have a set structure or should it be a kind of semantic network? In the first case, elements[3] in the memory are not related to each other

---

[2] The notion of a production system was in more general terms introduced in Chapter 2.

[3] An element of a declarative memory in PRISM is a list, for example: (this is an element), (Goal: succeed), ((also) (an) ((element))).

besides sharing the same data space. In the second case, elements are linked to each other so that they can be reached by spread-of-activation mechanisms. Further decisions concern quantitative attributes for elements. Should elements have a quantitative attribute, such as recency: a time tag indicating at what cycle the element was stored into the memory? Such attributes can be used, for instance, to simulate forgetting.

Similarly, the designer has to decide how many procedural memories he needs and decide on the declarative memory/memories a procedural memory will match against. He also has to decide on the conflict resolution scheme: If more than one production matches against elements in working memory in a given cycle, should they all fire or should there be a selection procedure resolving the conflict? In case conflict resolution should be selective, one or more quantitative comparison dimensions have to be defined for the productions, and an ordering scheme for those dimensions has to be specified. For example, production rules which have a *strength* parameter attached can be ordered according to strength and the one(s) with the highest strength value can be selected to fire.

Further, for both declarative and procedural memories, automatic tests on quantitative attributes of elements or productions, respectively, can be defined that become activated in each cycle. For example, it could be tested in each cycle whether elements in a declarative memory fall below a certain recency threshold, and those elements could be removed.

With these decision points in mind, let us turn to the definition of HDD-SH's architecture.

Memories

HDD-SH comprises four memories. Two of them contain rules (procedural memories), two contain declarative information about experiments and other task related information. The procedural memory *CONTROL* holds the system's knowledge of how to acquire knowledge, whereas the knowledge itself (hypotheses in rule form) is stored in the second procedural memory, *Hypothesis Memory (HMEM)*. The declarative memory *Environment (ENV)* stores all the information available from the task environment, whereas the second declarative memory, *Working Memory (WM)*, holds information which is in the system's current focus of attention. Figure 6.2 depicts the relations between these memories.

ENV is a declarative memory holding complete information about all experiments encountered in the course of learning. It thus constitutes an analog to an external information storage medium, REFRACT's *NoteBook*. ENV is internally structured as a network with a spread-of-activation mechanism. However, no hypotheses about selective information retrieval are currently implemented; when the spread-of-activation mechanism for ENV is activated, it simply retrieves all information related to a given experiment. In other words, each time ENV's spread-of-activation mechanisms is evoked, a 'row' from the *NoteBook*, describing the quantitative aspects of one experiment, is transferred into Working Memory.

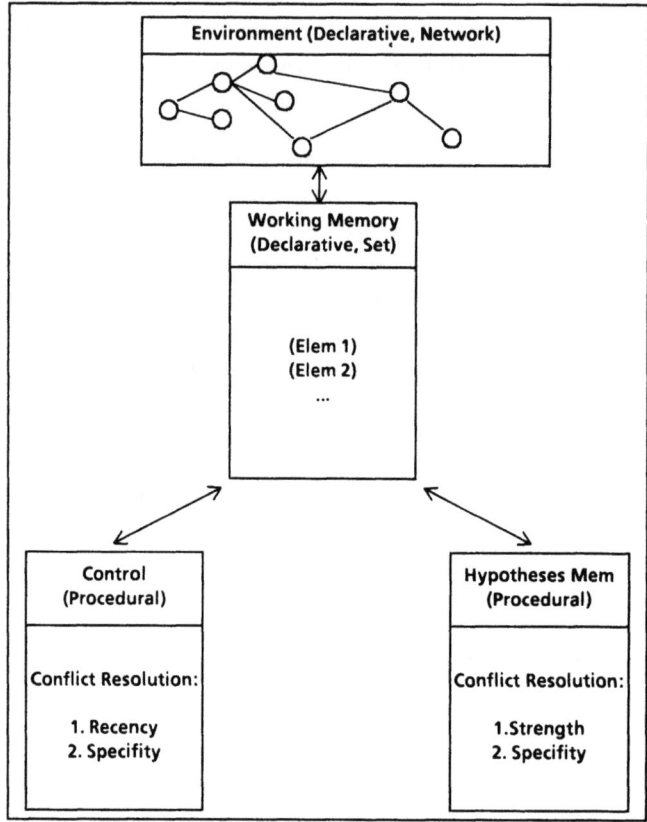

Fig. 6.2: Production system architecture of HDD-SH

WM is a declarative memory (with a set structure) that corresponds to human short term memory. It can hold only a limited amount of information, which is assumed to be in the learners focus of attention. Specifically, WM can hold information about a maximum of two experiments at a time. This is the minimal amount of information required for the system to apply heuristics about trends. In addition to information about experiments, WM holds a constant number of information pertaining to the kind of variables available in REFRACT: whether they are dependent or independent, nominal or numerical, and so on. Elements stored in WM have a quantitative attribute, *recency*, attached. The recency value of an element is equal to the value of a counter which is increased by 1 whenever an element is added to WM. Thus the element with the highest recency value has entered WM last. WM interacts with ENV: When the system moves its focus to another experiment while two experiments reside in WM already, one of the two experiments is removed before the new one is read into WM (see Figure 6.3).

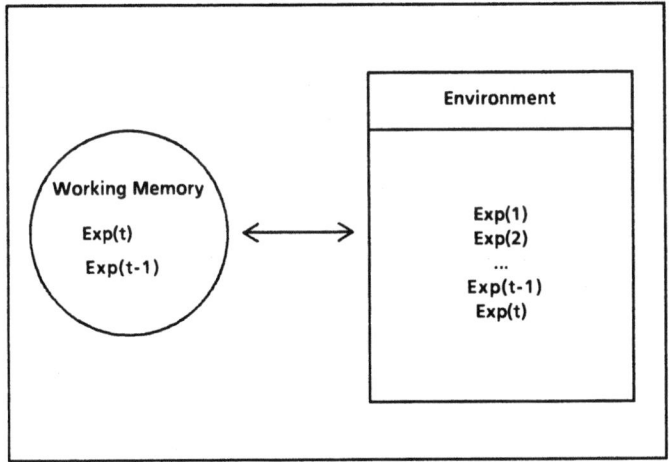

Fig. 6.3: The interaction between memory WM and ENV

Knowledge of how to construct hypotheses, how to store and how to apply them is captured in production rules which are stored in the procedural memory *CONTROL*. It matches against WM. CONTROL contains rules that organize the program's goal sequence (see Figure 6.1). Rules in this memory 'know', for example, when to trigger the construction of a design, when to make a prediction, or what to do when a prediction was wrong. Furthermore, CONTROL holds rules necessary to find trends in data and to induce equations. The following conflict resolution scheme is defined for this memory. Instantiations of productions[4] are first ordered according to the recency value of the element in WM against which their first condition matches.[5] Since all productions in CONTROL (see below) have as their first condition a pattern describing a goal, such as *(GOAL: generate design)*, instantiations become ordered according to the recency of their goal pattern. Next, the production (or productions) matching against the element with the highest recency value is selected. This step assures that the system selects only those instantiations that deal with the most recent goal. If there is more than one instantiation in the selection set, they are ordered again, this time by adding together the recency values of all elements matched by an instantiation. A single instantiation is selected from the set of those instantiations that have the highest value on this cumulative recency measure. After this second selection step, the production that will finally fire is the most specific one from those that match, where specifity is measured in terms of the number of pattern-variables appearing in the condition.

---

[4] In order to understand conflict resolution in PRISM, we have to acknowledge the distinction between a production and its instantiations. A single production can match in more than one way against the elements in a declarative memory, resulting in different instantiations. For example, if a production p has a single condition (Goal: =any), and there are two goal elements stored in the matching declarative memory: (Goal: Go) and (Goal: Stop), two instantiations of p are created: in $p^i{}_1$, =any is unified with Go, in $p^i{}_2$ with Stop. Conflict resolution always works with instantiations of productions.

[5] Conflict resolution can be based on quantitative attributes of the production rules themselves (production strength, for instance), or on attributes of the elements they match against (for example, their recency).

Hypothesis-rules are stored in *Hypotheses Memory* (HMEM). As long as they reside in this memory, they are available for prediction purposes. Initially, HMEM is empty (more precisely, it contains a single rule which says that nothing can be predicted). While HDD-SH works its way through a series of experiments, it acquires hypothesis-rules and stores them into HMEM in form of production rules. HMEM matches against WM. Conflict resolution in HMEM has the following meaning: In case more than one production (= hypothesis) is applicable given the current working memory content (= description of the current design, and some other information), which hypothesis/es should fire (= assert a prediction)? Two quantitative attributes of productions in HMEM are defined in all HDD versions: *strength* and *specifity* (see Figure 6.4).

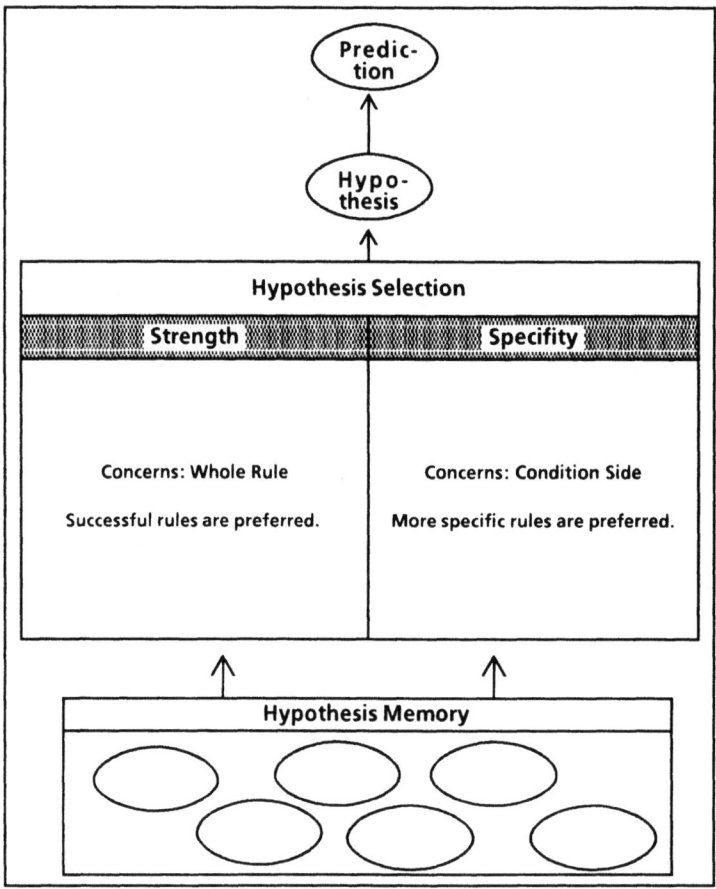

Fig. 6.4: Hypothesis selection during the prediction step

The strength of a hypothesis-rule is a measure of its success, as explained in Chapter 5. Specifity, the second attribute used, is calculated by counting the number of strings appearing in the condition part of a hypothesis-rule.[6] In general, rules with the highest strength value and the highest specifity value are preferred over rules with lower values on these dimensions. The details of conflict resolution in HMEM will vary across different HDD implementation variants and this will of course imply a change in the architecture.

A final feature of HMEM is that the hypothesis-rules stored in it are subject to *forgetting*. Whenever the strength parameter of a rule falls below a certain threshold it is automatically removed from HMEM. In order to enter HMEM again, the hypothesis has to be generated again by means of HDD-SH's phenomenon creation operators.

Learning Mechanisms

The discrimination learning process, including strength tuning, is another major aspect of the system's architecture. It belongs to the architecture and not to the system's knowledge since learning is not encoded in terms of productions, but as LISP functions which cannot be inspected from the 'knowledge level'. In other words, discrimination learning does happen *automatically*, only the decision *when* and *what* to discriminate is under the system's control.

The principles behind discrimination learning have been described in Chapter 5. The details of its implementation will not be described in this thesis. The basics can be found in Langley & Neches (1981). Plötzner & Opwis (1987) describe in detail an implementation for the InterLisp version of PRISM, which is used in my simulations with some minor modifications. I will confine myself here to mentioning the main functions and the parameters used in the HDD simulations.

The discrimination algorithm is triggered by a call to the function DISCRIMINATE which takes as argument a proposition from working memory. In case of HDD-SH, this function is called when a erroneous prediction was made. Discriminating a production rule requires that information about its last successful application is stored with the rule. (Information about the failure context needs not to be stored since it still resides in WM when the discrimination process is triggered.) In general, the function ASSIGN-CREDIT is provided in PRISM to attach a working memory context to a production rule. ASSIGN-CREDIT is used to link a prediction with its hypothesis. Finally, two functions are provided to change the strength parameter of a rule: INCREASE-STRENGTH and DECREASE-STRENGTH.

The following parameters are relevant to specify the discrimination process in HDD-SH. The *initial* strength of a rule when it gets inserted into HMEM is 1.0. In case a rule gets *rebuilt* (i.e., a

---

[6] For conflict resolution purposes, the specifity dimension could be introduced indirectly by using the recency measure of elements in WM the hypothesis rule matches against. However, I prefer an explicit representation of a hypothesis rule's specifity so that it can be manipulated directly, if necessary, and to make it possible to build summative HMEM statistics.

production is generated either from the equation finding algorithm or during discrimination which is identical to a production already residing in HMEM), its strength parameter is increased by 1.0. The function INCREASE-STRENGTH will result in an *increase* of a rule's strength value by 1.0; *decreasing* a rule means to multiply the current value by .25, i.e., strength is reduced to a quarter of its original value. Thus, 'punishing' a rule results in a much more drastic strength change than 'reinforcing' a rule (cf. Anderson, 1983); failure is treated as a more serious event than success. Finally, it needs to be mentioned that the discrimination algorithm considers the context a rule was created in as its first 'good' context. This way, discrimination can be triggered the very first time a hypothesis-rule leads to a wrong prediction.

Representation Language

Information about experiments is represented in the production system in form of feature vectors. Table 6.1 displays the variables and the values they can take.

Table 6.1

The numerical description language for experiments

Attribute	Value-set
Substance	(Glass, Flint, Diamond)
Shape	(plane,100,120,140,-100,-120,-140)
ODist	(-250,-200,-150,-100,-50)
Alpha	(-20,-15,-10,10,15,20)
$Theta_1$	[any real number]
$Theta_2$	[any real number]
Gamma	[any real number]
IDist	[any real number]

The variable *Beta* is not included in the simulation programs. It is part of REFRACT's variable set, but it is completely irrelevant and was never used by any of my subjects. Including it only would have added more noise to the simulations without changing anything of importance.

Declarative Domain Knowledge

The knowledge all HDD models - so also HDD-SH - have available in declarative form (i.e., stored in WM and ENV) is displayed in Table 6.2

Table 6.2

Declarative information stored in memory WM
___

Information on Variable Type and Role:

    (Independent Radius)
        ...and so on for ODist, Alpha, Theta1.
    (Dependent IDist)
        ...and so on for Theta2, Gamma.
    (Nominal Substance)
    (Numeric Radius)
        ...and so on for ODist, Alpha, Theta1, IDist, Theta2, Gamma.
    (Angle Alpha)
        ...and so on for Theta1, Theta2, Gamma.
    (Distance ODist)
        ...and so on for Radius, IDist.
    (SameType (Radius IDist))
        ...and so on for (ODist, IDist), (Alpha, Theta2), (Theta1, Theta2), (Theta1, Gamma), (Alpha, Gamma).

Information on Variable Values:

    (Values Substance glass flint diamond)
        ...and so on according to table 4.1

Information about Focus of Attention and Learning History

    (CurrentExperiment <c>)
    (ExpsInFocus:(<e_i> <e_k>))
    (ExpsInNoteBook: <e_1, e_2, ...>)
    (Last experiment <e_1>)
    (FailureCount <c>)
___

Procedural Knowledge

The procedural knowledge of HDD-SH can be classified into four functional groups: rules for (1) experiment design, (2) prediction making and evaluation, (3) hypothesis evaluation and modification, and (4) hypothesis generation.

Rules for Designing an Experiment. As mentioned when developing the conceptual background, this step is not accomplished by my computer models. HDD-SH and the other variants discussed later can not design experiments on their own. Instead, they accept design descriptions provided by the programmer as input. HDD-SH as described in this section will use data from a factorial design scheme. *Alpha* is varied first (with two values: 10 and 15), then *ODist* (-100, -200), Radius (*plane*, 100), and finally substance type (*glass, flintglass, diamond*). This results in an sequence of

24 experiments.

Rules for Making a Prediction and Evaluating the Prediction. Finding a prediction is done after information about the current design entered WM. Constructing a prediction is accomplished by activating[7] memory HMEM and by letting a hypothesis-rule from HMEM fire. This will result in having a prediction placed into WM. For example, let us assume that the following hypothesis-rule fires:

```
(Hypothesis-0025
    ((Goal: predict)
    (Exp =e Substance =iv1)
    (Exp =e Radius =iv2)
    (Exp =e ODist =iv3)
    (Exp =e Alpha =iv4)
    (Exp =e Theta1 =iv5))
    -->
    (($ADD-TO WM
    ($ASSIGN-CREDIT ((&GENSYM Prediction-) ISA: Prediction
    DV: IDist IV: ODist Form: (LAMBDA (IV) (TIMES IV 2.0))
    Value: (&APPLY (LAMBDA (IV) (TIMES IV 2.0)) =iv3))))[8].
```

It says: *For all possible combinations of design variables, predict that the dependent variable IDist will be equal to two times the independent variable ODist.* In a context where the condition element with the pattern-variable =iv3 matches against the following working memory element,

(Exp exp-3 ODist -200),

=iv3 gets bound to -200. In this case, the following prediction is placed into WM:

```
(Prediction-0031 ISA: Prediction
    DV: IDist IV: ODist Form: (LAMBDA (IV) (TIMES IV 2.0))
    Value: -400)
```

When *ODist* takes the value -200, say, the resulting value for the prediction is IDist = -400. *Prediction-0031* is linked internally to *Hypothesis-0025*, where it is derived from, by means of the PRISM function ASSIGN-CREDIT. This linkage will be needed later when it has been determined whether *Prediction-0031* was correct or wrong and we need to strengthen or weaken the responsible hypothesis.

When no hypothesis can be found that matches the design description in WM, no prediction is derived.

---

[7] In PRISM, a procedural memory can be pacified--its productions are then not considered during the matching phase--and it can be reactivated.

[8] This is an example for an actual PRISM rule with the following syntax: (<rule name> (<condition part>) --> (<action part>)). The dollar sign ($) precedes a PRISM memory command, the ampersand (&) indicates a LISP function call. Strings that start with an equation sign (=) are treated as pattern variables by the interpreter. They can match against any expression.

In order to find out whether its prediction was right or wrong, the system needs to be provided with feedback information, that is, with the correct values for *Theta2*, *Gamma* and *IDist*. These values are provided as additional input to the program. It then compares the predicted value with the correct value and inserts a message into WM saying that the prediction was correct or wrong.

Rules for Evaluating and Modifying a Hypothesis. In case the prediction was correct, the following hypothesis evaluation rule fires:

```
ModifyH.CorrectPred
IF      the goal is to evaluate the hypothesis,
        and the prediction was correct
THEN    increase the strength value of the hypothesis,
        and save the application context as an example of a positive
        application of the hypothesis.
```

It increases the strength value of the hypothesis behind the correct prediction and links the current WM content to that hypothesis, indicating that this was a design where the hypothesis led to a correct prediction. On the other hand, if the prediction was wrong, this rule will fire:

```
ModifyH.WrongPred
IF      the goal is to evaluate the hypothesis,
        and the prediction was wrong
THEN    decrease the strength value of the hypothesis,
        and call the discrimination process on the hypothesis.
```

The strength parameter of the rule is decreased and the discrimination procedure is called, using the current WM content as an example for a situation where the hypothesis did make a false prediction.

Rules for Creating Phenomenon Descriptions (Equations). When no prediction could be derived or when the prediction was false, the system attempts to find new hypotheses in the sense that it looks for new relations in the data recently accumulated. As described in Chapter 5, a two step algorithm is employed that first discovers trends and then equations. The rules operative were also described in Chapter 5.

Miscellaneous Rules. Two rules manage the transfer of elements between the working memory WM and the external memory ENV (see Appendix VI.1).

Table 6.3 summarizes the most important features of HDD-SH. To complete the description of HDD-SH's structure, its flow of control is sketched out in Figure 6.5.

Table 6.3

Architecture for learner model HDD-SH

PREDICTION MAKING	
Conflict Resolution Scheme	Select_One_Best(Strength)
Forgetting in HMEM	Strength < .5
HYPOTHESIS MODIFICATION	
Learning Method	Discrimination
Strength Revision	
Increase	PLUS 1.0
Decrease	TIMES .25
HYPOTHESIS GENERATION	
Initial Strength	1.0
Increase When Rebuilt	1.0
Trend Detectors	Directly Proportional, Inversely Proportional
Mathematical Operators	PLUS, DIFFERENCE, TIMES, QUOTIENT

6.2.2 Learning Results and Learning Process

Learning Results

In the course of working its way through the 24 experiments of the factorial design, HDD-SH generated 59 equations as general hypotheses, of which four were further specialized, one of them two times. Overall, 77 hypotheses were generated. One hypothesis was 'forgotten' by the system since its strength parameter sank below the threshold value. At the end of the experiment sequence, HDD-SH found the correct equations, but also many wrong and spurious ones. Most of the hypothesis-rules did not acquire a strength value above the initial one (1.0), mostly because they were never used for prediction making. Table 6.4 shows how many hypotheses were created based on discrimination and Figure 6.6 displays the hierarchy of rules resulting from the discrimination process. As a final overview of the learning results from HDD-SH, I list all those hypotheses that have a strength value greater or equal 2.0, i.e., that were strengthened at least once (Table 6.5).

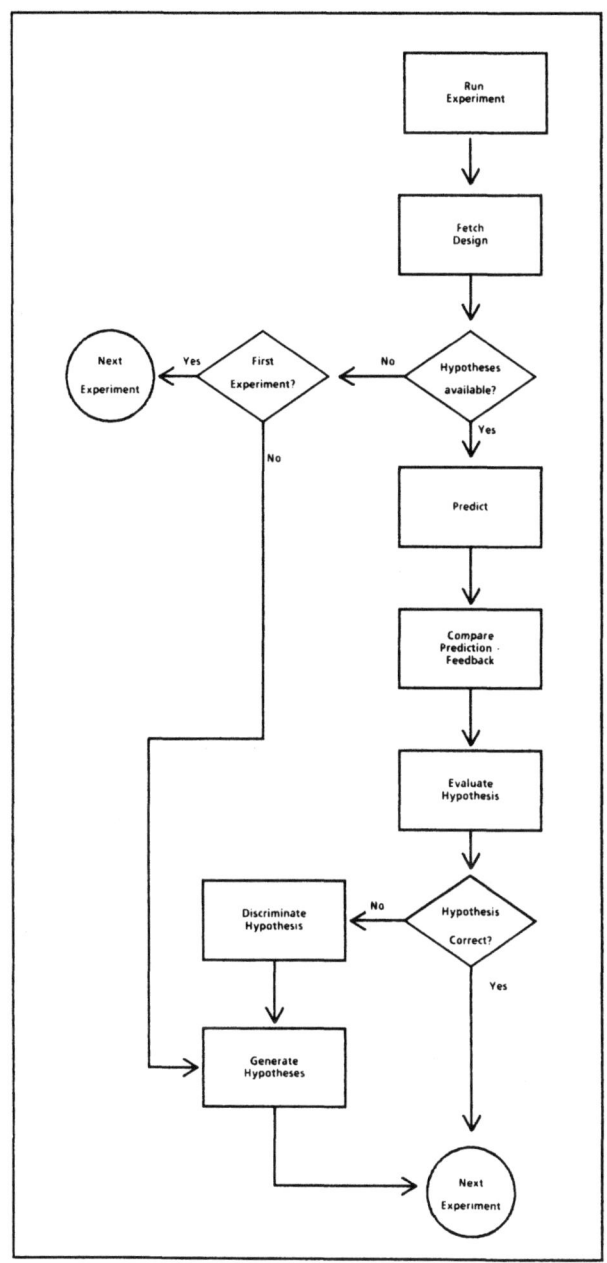

Fig. 6.5: Flow of control in HDD-SH

Table 6.4

Distribution of number of conditions

No. of Conditions	No. of Hypotheses	%
0	59	76.6
1	15	19.5
2	3	3.9

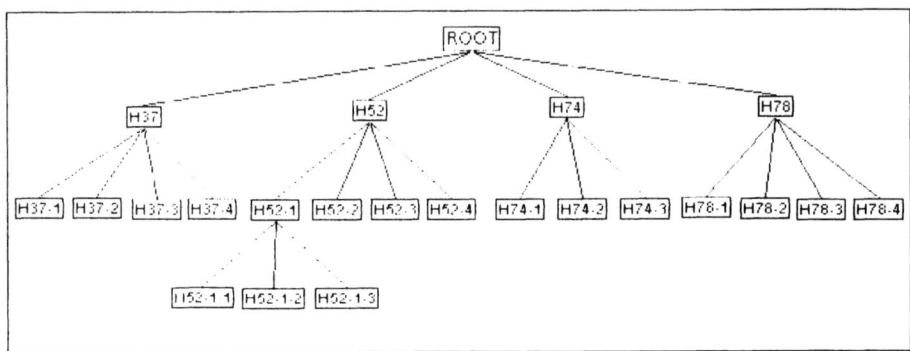

Fig.6.6: Discrimination tree for HDD-SH. Only those hypotheses are displayed that became discriminated. A similar tree is created online by the programming environment HDD-SH runs in.

Table 6.5

Final hypotheses with strength greater or equal 2.0

Hypothesis	Condition	Equation	Strength
H-31		Gamma=Alpha/2.5	2.0
H-37-1	Alpha 15	Theta2=Theta1*-.4	2.0
H-37-2	ODist -200	"	2.0
H-37-3	Radius -100	"	2.0
H-37-4[a]	Subst. glass	"	3.0
H-52-2	ODist -200	Theta2=Theta1*-.33	2.0
H-52-3	Radius 100	"	2.0
H-52-4[a]	Subst. flint	"	2.0

Note. The hypothesis labels do not indicate the actual number of hypotheses. For instance, H-0033 does not mean that this was the 33rd hypothesis generated. What can be concluded is sequence; for example, H-0031 was generated before H-0037-1.

[a]Correct

For substance type *glass* and *flintglass*, the correct equations have been found (H-37-4 and H-52-4). For *diamond*, the correct phenomenon part was generated, and so was the correct

does not show up in this listing. We see that this hypothesis set contains still rules with an incorrect phenomenon part (H-31) and with incorrect conditions (H-37-1 and -3, H-52-2 and -3).

Learning Process

In order to describe the learning behavior of HDD-SH, I provide two kinds of traces that can be found in Appendix VI.2 and VI.3, respectively. The first trace illustrates the goal sequence of HDD-SH by following the system through the first three experiments. This trace is documented in-line and not further explained here. The second trace (Appendix VI.3) shows the learning steps of HDD-SH through all 24 experiments (plus three test items). Let me summarize this trace.

Exp-1. No prediction is made since no hypothesis exists. Further, no hypotheses are generated since there is not enough information available; recall that the trend detectors require data from two experiments.

Exp-2. A first set of eight hypotheses is generated. Among them H-0037 which covers the relation between *Theta2* and *Theta1* correctly for medium type glass, but which is in this form too general.

Exp-3. For the first time, a prediction is made. Since all of the eight initial hypotheses generated in the last step have the same strength parameter, one of them is selected randomly. Luckily, its the right one, H-0037. It leads to a correct prediction and its strength value is increased.

Exp-4 to Exp-8. H-0037 is repeatedly applied and leads always to the correct prediction. It therefore acquires more and more strength.

Exp-9. Here substance type changes from *glass* to *flintglass*, and the overly general H-0037 leads to a wrong prediction. The discriminator is called and it generates four specializations of H-0037. Among them the correct hypothesis for substance type *glass*, H-0037-4. In addition, ten new hypotheses are built as a result of looking for functional relations between Exp-8 and Exp-9.

Exp-10. In this experiment, H-0037 is used again for prediction making, since it has still enough strength to dominate competition with the other hypothesis-rules. It turns out to be wrong again, since substance type is still *flint*. This time, strength-reduction pushes H-0037's strength value under the threshold of .5 and it is removed from memory HMEM. Its specializations, however, are rebuilt and acquire therefore additional strength. Seven hypotheses are newly built, among them one that was already built in Exp-9: H-0052; it is rebuilt. It captures correctly, though overly general, the relation between *Theta1* and *Theta2* for substance type *flintglass*. Rebuilding results in a strength increase for H-0052.

Exp-11 to Exp-16. H-0052 dominates the scene since it makes correct predictions for all experiments with substance type *flintglass*.

Exp-17. Here substance type switches to *diamond*, and H-0052 turns out to be wrong for this case. Four specializations are created, among them the correct one for substance *flintglass*, H-0052-4. Further, new general hypotheses are generated, including the correct one for substance type *diamond*, H-0078.

Exp-18. The wrong hypothesis used on Exp-17 is used again since it still has the highest strength value. The specializations built before are confirmed and acquire additional strength. In the process of looking for new functional relations, two relations are rediscovered, as captured in H-0074 and H-0078.

Exp-19 and Exp-20. From the two general hypotheses that have the highest strength value (2.0), one is selected randomly in Exp-19: H-0074. It leads to a correct prediction in Exp-19 and Exp-20, where surface form is *plane*. Since its phenomenon description is only true under this condition, it leads to a wrong prediction in Exp-21, where surface forms becomes *convex*.

Exp-21 to Exp-24. After Exp-19 was run, H-0078 acquired enough strength to dominate the rule competition and produces correct predictions for all experiments with substance type *diamond*.

Exp-25 to Exp-27. The last three experiments constitute test items and do not belong to the factorial design sequence. In the course of working on these items, the overly general H-0078 is discriminated. The correct rule for diamond is found in Exp-25 in form of H-0078-4.

The plot in Figure 6.7 summarizes the hypotheses generation process on a coarse level. The number of hypotheses stored in memory HMEM is plotted against 'learning time' (a counter of production system cycles). 'Plateaus' in this plot correspond to stages where HDD-SH did not have to generate new functions since it made correct predictions. The graph in Figure 6.8 shows the strain placed on HDD-SH's working memory. The number of forty elements is chosen as baseline since this is the number of elements that stay constantly in WM. The peaks result from elements in WM describing trend and function information.

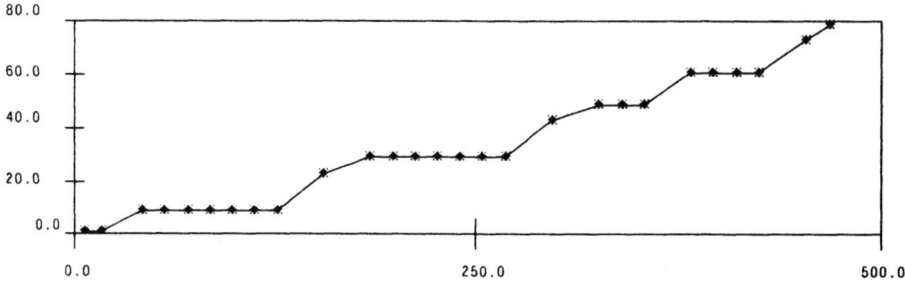

Fig.6.7: Development of number of hypothesis-rules over production system cycles (HDD-SH)

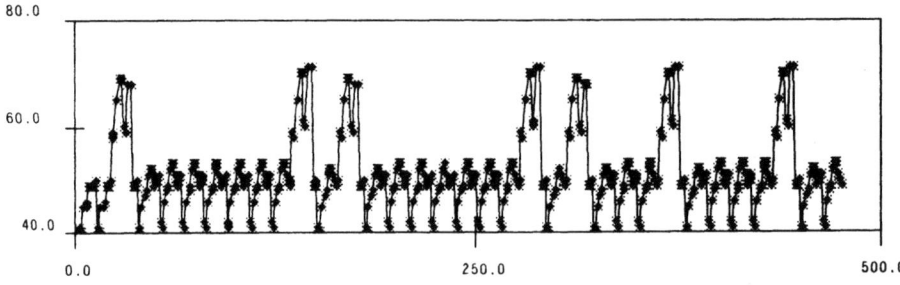

Fig.6.8: Development of number of elements in working memory over production system cycles

What is accomplished with this first implementation? We have a rule-based computational model that can solve the induction problem in REFRACT, given that it receives a sufficient number of experiments. In particular, the three main problems of induction are considered (Holland et al., 1986): (a) generation of rules that can extract regularities from the environment (here: from the optics simulation) is realized as function induction; (b) refinement of knowledge and performance based on feedback from the environment is established by means of discrimination learning; (c) organization of rules into clusters and default hierarchies is (partially) accomplished as an side-effect of discrimination learning: since specific rules do not immediately replace their general predecessors but enter competition with them, the system can fall back on these more general rules when the specific rules do not apply.

The main problem of a discovery learner that works as HDD-SH does is that he generates too many functions in the hypothesis generation step and generates too many specializations in the hypothesis modification step. In the next section, I turn to the question of how to constrain the model.

6.3 Constraining HDD-SH

In principle, hypothesis generation and selection in HDD-SH can be further constrained in two ways: Restrictions can be put on the hypothesis generation (function induction) step, or on the hypothesis modification step. Restricting the hypothesis modification step would mean to constrain the discrimination learning process, for example, by changing the triggering conditions (e.g., not discriminating after every wrong prediction, but only after k wrong predictions in sequence) or limiting the number of discriminated hypothesis-rules the algorithm can output. However, since discrimination learning is such a general learning mechanism I do not want to put task-specific constraints on it, at least not as the first choice. I consider modification of the - already more domain-specific - function generator as the better candidate for implementing additional constraints.

In order to limit the function-induction component of HDD-SH, two options arise. For one, the hypothesis creation and modification operators can be made more 'intelligent'; in particular, equation generation could be improved by considering more data points that an equation must account for; or equation generation could be coupled more closely to the result of the prediction-feedback comparison; or the number of function induction operators could simply be restricted. Secondly, instead of making the hypothesis generator itself more intelligent, tests can be attached to it that filter out the more plausible candidates.

Filters on hypotheses could easily be implemented into the existing HDD version: as preference criteria in the prediction step, that is, as part of the conflict resolution process for HMEM and therefore as part of the architecture. Constraints of this kind constitute strong assumptions in as much as the claim is that the learning system--human or artificial--will *automatically* select hypotheses that are more specific, less complex, etc. This is to be contrasted with constraints on function induction that originate from a learners *deliberate decision* not to look for new hypotheses given a prediction error, or where a learner does not have the *knowledge* about how to find functions in data, etc.

One possible filter is *specificity*. A desirable feature of a learning system that is based on discrimination is that it prefers specific rules over more general ones. This enables the system to consider rules for application that cannot acquire a lot of strength since they describe situations in the environment which do occur less frequently. A system that has at its command strong rules that describe common situations and default behavior and that is at the same time able to react to specific situations by calling on specific pieces of knowledge can react flexibly and adaptively (see the notion of a 'default hierarchy' in Holland et al., 1986).

While it is in general true that specific hypotheses should get a chance to acquire strength, it presupposes that their action parts are of a kind so that they make constructive contributions to the system's performance. If this is not the case, if the action parts of many hypothesis-rules are weak, the depth-first search for conditions has disastrous effects. Since more specific hypotheses are preferred over more general ones, the system will quickly generate many highly specific rules. Since HDD-SH generates many hypotheses containing spurious relations between variables, it runs into the problem that, using a depth-first search for conditions, it is for the most time occupied with finding additional conditions on often spurious functional relations. We can conclude that the depth-first search for conditions can not be successful in REFRACT. It does only work if the generator for phenomenon descriptions is very selective and correct.

A way specificity can be more safely employed in HDD implementations is to combine it with the strength selection in the sense that first the rules with the maximal strength are selected and within this set the one(s) with maximal specificity are selected. This scheme is employed in the next version of HDD, discussed below. Before coming to this second version, let me first consider another filter criterion, *simplicity*.

Simplicity/complexity of a hypothesis refers to the *phenomenon* part of a rule, not, as with specifity, to the condition part. The problem is that it is hard to say what should count as a 'simple' phenomenon and what as a complex one. Note that a simplicity measure is indirectly implemented in HDD-SH by means of the function induction operators. As described in the last chapter, these operators do already limit the set of possible functions to be induced considerably by restricting them to the simple *<DV>=<IV><Operator><Constant>* form. This simplicity criterion is not realized as a filter, but as a constructive constraint, a procedural bias. Since to my knowledge generally useful evaluation functions in the context of equation induction have not been formulated yet, I did not further consider filter functions relating to the content of an equation.

That leaves us with the option to restrict the function generation step by making the generator itself more 'intelligent'. Note that this step from filtering out implausible hypotheses to not constructing implausible hypotheses (or at least not as many) means to prefer *process constraints* over *product constraints*. As discussed in Chapter 2, process constraints hold the promise to be more generally applicable than product constraints. With the next two models, I will introduce two process constraints on function induction, one based on the example of BACON.1, the other one based on a strategy employed by a successful subject in the empirical study.

6.3.1 Restricting Function Induction By Preferring Operators: HDD-SHR

In this section, a variant of HDD is developed where the hypothesis generation step is improved. This more constrained function inducer does use only two (instead of four) functional operators to generate equations, similar to BACON.1, that builds only equations using the product operator (for inversely proportional trends) and the quotient operator (for directly proportional trends). I employ this constrained function induction algorithm in a further version of HDD-SH: HDD-SHR(estricted).

Conflict resolution in HMEM prefers again (as in HDD-SH) hypotheses with high strength value over those with low strength. Thus, HDD-SHR performs a best-first search for conditions. Two additional modifications are built into HDD-SHR. New is that if there's more than one hypothesis with maximal strength value then the one with the highest specifity rating is preferred. Specifity is operationalized as the number of constants appearing in a rule's condition. Second, HDD-SHR incorporates a limit on the number of conditions a hypothesis can have at most (more precisely, on the number of constants that replace pattern variables in the condition side): A hypothesis can be discriminated at most two times. The rational behind this constraint is that the system should not create hypotheses with an overly complex condition side, but rather attempt to find better phenomenon descriptions. The changes from version HDD-SHS to HDD-SHR are highlighted in Table 6.6.

Table 6.6

Architecture for learner model HDD-SHR

PREDICTION MAKING	
Conflict Resolution Scheme	**Select_All-Best(Strength)**
	Select_One_Best(Specifity)
Forgetting in HMEM	Strength < .5
HYPOTHESIS MODIFICATION	
Learning Method	Discrimination
Limit on no. of conditions	**LEQ 2**
Strength Revision	
Increase	PLUS 1.0
Decrease	TIMES .25
HYPOTHESIS GENERATION	
Initial Strength	1.0
Increase When Rebuilt	1.0
Trend Detectors	Directly Proportional,
	Inversely Proportional
Mathematical Operators	**TIMES, QUOTIENT**

The restriction of the function generator does result in an improvement of model performance. In a run with the same experiment sequence as used with HDD-SH, HDD-SHR created 55 hypotheses, instead of the 77 produced by HDD-SH. Of these, 32 are general unconditional hypotheses, 6 of them were discriminated one time. It found the correct rules for glass and flintglass; the correct hypothesis for substance type *diamond* was created but did not acquire the necessary strength value yet. Also, many of the hypotheses do incorporate an irrelevant condition element (specific values for radius, *Alpha*, *ODist*), and many do use *Gamma* as the dependent variable. While all of these hypotheses cover more than one experiment correctly, they focus on only partially relevant variables.

6.3.2 Restricting Function Induction By Preferring Operators and Variables: HDD-SHLR

The particular modification suggested now attempts to integrate an observation made on Subject S10. He at times generated a new equation by focusing on the variables and the function included in his current (wrong) prediction (see Chapter 4.4). That is, instead of looking for all relations between dependent and independent variables after a wrong prediction, guided by a proportionality heuristic, he concentrated on the single one relation that was the content of his prediction. In terms of search, this is a heuristics that corresponds to using the Channel *e* in the GRI model (cf. Fig 1.1): the result of the hypothesis testing step (i.e., the comparison prediction - feedback) is not only used to purge the hypothesis when it turns out to be wrong, but influences also the generation of 'better' hypotheses. The variant of HDD that incorporates this strategy is called HDD-SH*L(ocal)R(elation)*. Compared to HDD-SHR, this model employs a stronger restriction since it limits - at least temporarily - creation of new functions to one specific mathematical operator and two specific variables.

The approach is easily integrated in HDD. Starting point for the modifications is HDD-SH. In order to integrate the preference heuristic into HDD-SH, two new rules are added to the old program. The first one is a variant of *ModifyH.WrongPred*. Recall that this is the production in memory CONTROL that is triggered when a prediction is wrong and as its action part calls on the discrimination procedure. As stated before, it looks like this:

```
ModifyH.WrongPred
IF      the goal is to evaluate the hypothesis,
        and the prediction was wrong
THEN    decrease the strength value of the hypothesis,
        and call the discrimination process on the hypothesis.
```

The new rule that competes from now on with *ModifyH.WrongPred* is:

```
ModifyH.WrongPred.1
IF      the goal is to evaluate the hypothesis,
        and the prediction was wrong
        and there were less than three wrong predictions made in sequence
THEN    decrease the strength value of the hypothesis,
        and call the discrimination process on the hypothesis,
        and set the goal to generate new hypotheses,
        and set the goal to look for the local relation between the
        dependent and the independent variable incorporated in the
        hypothesis.
```

In other words, when the system did not derive continuously wrong predictions in the last two experiments, this production will cause the system not to look for new hypotheses by means of the find-trends-then-functions procedure, but to look for the current 'local' relation between the dependent and independent variable in the current hypothesis. As a result, only a single new hypothesis will be generated (or none if it already exists).

The second new rule is part of the hypothesis generator and takes care of fulfilling the new goal, 'look for local relation':

```
GenerateHypotheses.LocalRelation
IF      the goal is to generate hypotheses,
        and the subgoal is to look for the local relation
        and the current prediction is based on a hypothesis with operator
        =o, dependent variable =d, and independent variable =i
THEN    calculate the value for constant c using =o, =d and =i.
```

This rule causes the system to skip all the consideration and trend detection steps and to directly consider a single function-operator for two variables. It will keep the variables and the operator from the current prediction and adapt the *<Constant>* term in *<DV>=<IV><Operator><Constant>* to the current experiment. After this is done, the next experiment is run.

The two new rules will only be activated when the system did not repeatedly make wrong predictions in the last experiments (the current parameter restricts it to two wrong predictions in sequence). This limit is introduced in order to avoid that the system endlessly tries to improve equations that include irrelevant variables and/or operators. In other words, I assume that a learner gives up improving a relation that does not lead to success and will widen his 'horizon' again to look for alternative relations.

Usage of the heuristic strategy as described reduces the number of newly generated equations. In a run with the factorial design, HDD-SHLR created only 13 unconditional hypotheses in 26 experiments, three of which were discriminated once, and of these three two became discriminated two times. Overall, the program generated only 28 hypotheses, of which two became deleted due to 'forgetting'. Compare this with the 77 hypotheses generated by HDD-SH, the same version of HDD, but without the two rules that focus the system on the last employed operator and variables. However, the hypothesis-rules HDD-SHLR creates tend to be overly specific. Since the system does not immediately turn away from a functional relation that leads a wrong prediction, additional conditions are attached quickly.

Looking back at the last two models, HDD-SHR and HDD-SHLR, two critical remarks are in place. For one, the 'focus on local relation' strategy used by HDD-SHLR is very domain-specific. It works only because the Law of Refraction as present in REFRACT can be found by adapting a single parameter, given that *Theta2* is expressed as the product (or quotient) of *Theta1* and the parameter (which reflects the different values for optical density in *glass*, *flintglass*, and *diamond*). Secondly, while the restriction of function operators as employed in HDD-SHR can be justified rationally - trained scientists would calculate product and quotient relation when confronted with inverse and direct proportionality, respectively - it is not psychologically convincing. Why should students not trained in data analysis not calculate sums and differences? And as I observed in my subjects, they do calculate sums and differences between certain variables in REFRACT. So, the decision to use only product and quotient operators in HDD-SHR is somewhat ad hoc. The question is how to introduce constraints on function induction that are less ad hoc and less domain-specific.

6.4 Learning variable and operator preferences: HDD-SHOP

In this section, I attempt to resolve both problems by providing an explanation for where the preference for certain operators and variables might come from. I propose that this preference is developed in the course of learning in REFRACT: A learner might not only acquire knowledge in form of hypotheses, but he can learn which operators for constructing hypotheses are useful and which are not. This argument is extended by claiming that not only function-operator preferences are learned in this manner, but also preferences for specific variables.

The idea is the following: Success or failure of a hypothesis is not only utilized in order to tune the strength value of the hypothesis (and to discriminate the hypothesis if necessary), but feedback is also used to learn about promising operators and variables. Tuning of a hypothesis' strength value is coupled to tuning of the rules that suggested the mathematical operator employed in the hypothesis, and tuning of the rules that suggested the variables which entered the equation. The assignment of credit and blame to a hypothesis is thereby extended so that the rules that were involved in building the hypothesis are also credited or blamed, respectively. Figure 6.9 illustrates the idea.

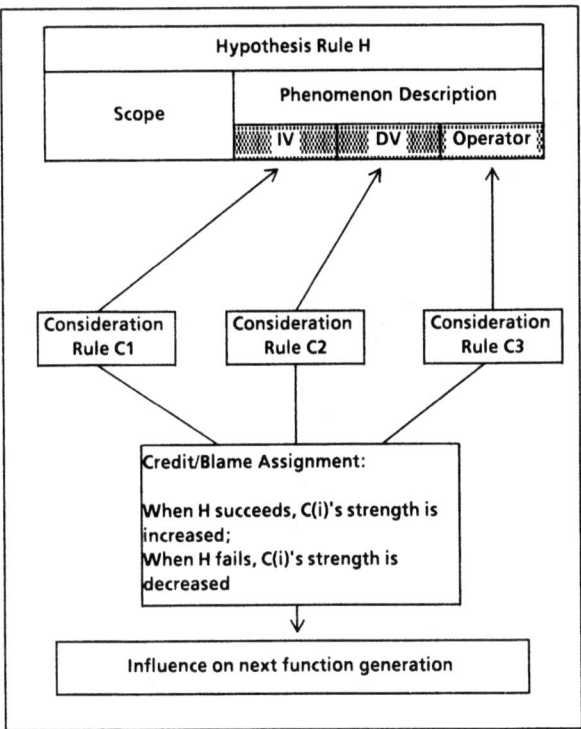

Fig. 6.9: The 'Bucket Brigade' mechanism

I coin this learning mechanism *bucket brigading*.[9] The name is borrowed from Holland et.al. (1986) where a similar, though much more elaborated scheme is proposed as the central credit/blame assignment mechanism in the context of so called *classifier systems*. The version of HDD using the bucket brigading idea is called HDD-SH*O(perator)P(reference)*.

In order to implement the bucket brigade mechanism in the HDD production system, we have to modify both the architecture and some of the rules responsible for modifying hypotheses. The architecture is changed as follows. A new memory, PHEN, is defined (Figure 6.10). It contains the rules responsible for trend detection and function induction that were so far stored in CONTROL.

---

[9] According to Webster's, "a bucket brigade is a chain of persons acting to put out a fire by passing buckets of water from hand to hand".

The trend detection rules are not changed. In addition to the trend-detection rules, there are now two sorts of rules responsible for the function induction task: *function-consideration rules* and *function-rules*. A function-consideration rule suggests to consider a specific function-rule under certain circumstances. For each of the four mathematical operators available for function creation there is a specific consideration rule. For example, the consideration rule suggesting to use the product operator is:

```
Consider.Product
IF     the goal is to find functions
       and a dependent and independent variable are related inversely
       proportional
THEN   suggest to consider the product operator for building an equation.
```

The product operator can only fire when its corresponding consideration has been placed into WM.

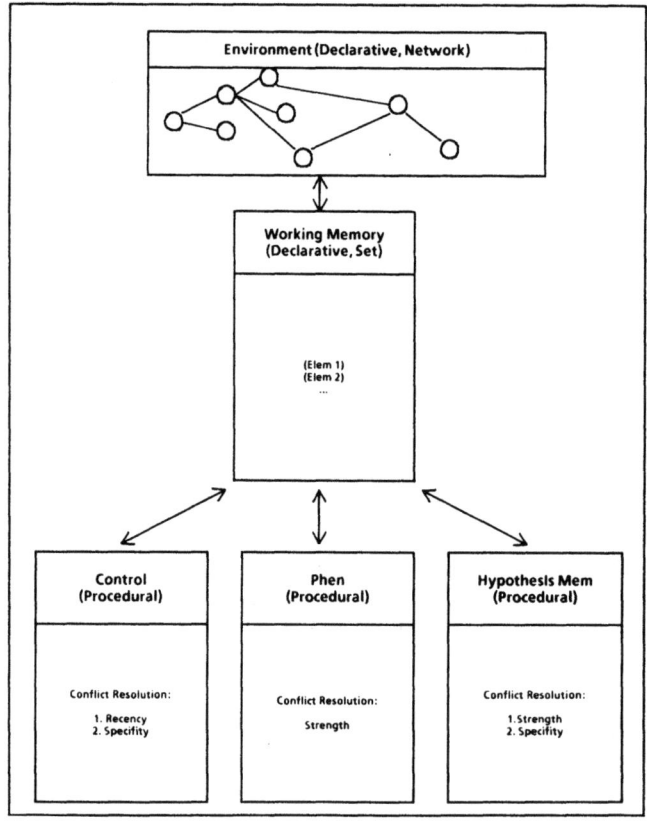

Fig. 6.10: Architecture of HDD-SHOP

In addition to function-consideration rules there are *variable-consideration rules*. For each of REFRACT's dependent and independent variables, a specific consideration-rule exists in memory PHEN. Such a variable-consideration rule will suggest to consider its variable for inclusion in the trend finding and function finding step. Only variables that are suggested in this way are considered by the trend detection rules and function induction rules.

By now we have four groups of rules stored in production memory PHEN: variable-consideration rules, function-consideration rules, trend detectors and function inducers. The advantages of this new knowledge decomposition will become clear when we consider PHEN's conflict resolution regimen and its interaction with the rules in memory CONTROL. Production rules in PHEN have a quantitative parameter *strength* attached. Conflict resolution for PHEN is defined so that its productions can fire in parallel; all of the instantiations with the maximum strength value are allowed to fire in the same cycle. The interaction of PHEN with memory CONTROL is depicted in Figure 6.11. The Figure shows the situation in an experiment after the hypothesis is evaluated as correct or wrong. At this point, evaluation of the hypothesis is coupled with evaluating the consideration rules involved in hypothesis construction. If a hypothesis turned out to be correct, an *error-counter* is set to zero. This counter is maintained in working memory. Further, the strength parameter of the successful hypothesis (*H*) is increased, and the strength parameters of the consideration rules involved in constructing *H* are increased.[10] For example, if *H* has the form

```
IF <any> THEN ODist = IDist TIMES 2.0,
```

the consideration-rules for *TIMES*, *ODist* and *IDist* have their strength parameter increased. If, on the other hand, the hypothesis *H* was wrong, the error-count is increased by one and the strength parameter of the hypothesis-rules as well as of the consideration-rules involved is decreased.

This tuning of the consideration-rules' strength parameter has effects when it comes to generate new hypotheses. The interaction between CONTROL and PHEN during hypothesis generation is organized as follows (Figure 6.12). When HDD-SHOP sets itself the goal to generate new hypotheses, a first subgoal is inserted in WM: *consider variables*. Then memory PHEN is called. In it, all variable-consideration rules have their condition part satisfied, since the general form of a variable-consideration rule is:

```
IF      (GOAL: consider variables)
        (Variable <x> not considered)
THEN    (Suggest to consider variable <x>)
```

---

[10] Technically, the consideration-rules are not stored with the hypothesis-rule, but a Lisp-function inspects the action side of the hypothesis and identifies the responsible consideration rules.

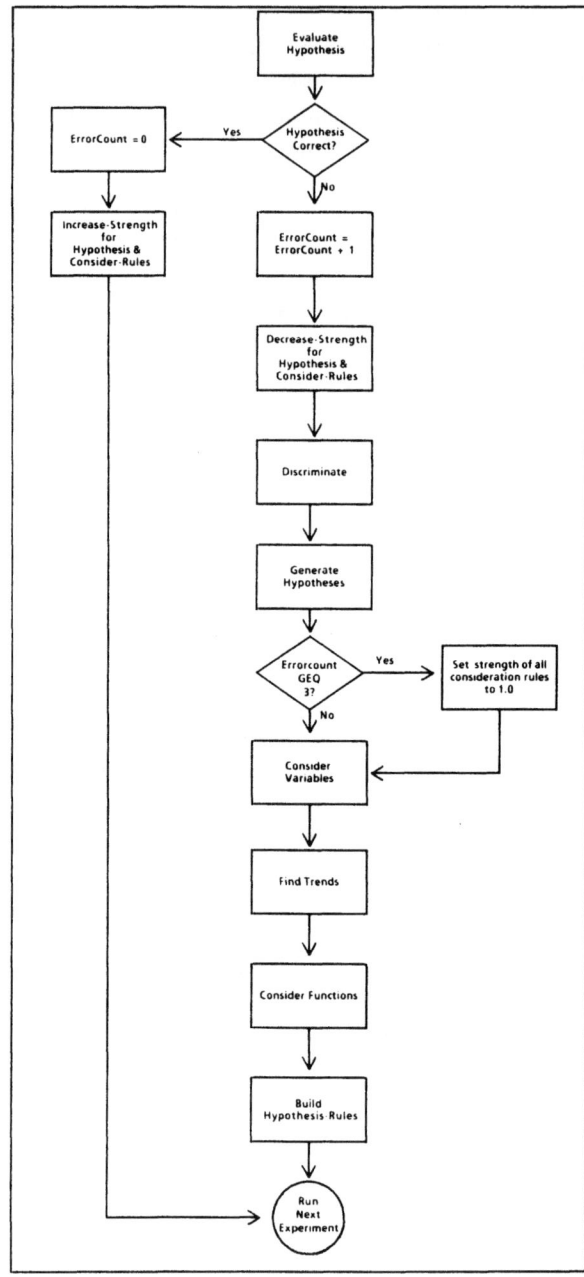

Fig. 6.11: Flow of control in HDD-SHOP

But only those variable-consideration rules with the highest strength value are allowed to fire. They insert their suggestions into working memory. Control is returned to CONTROL, which now sets the goal to *Find Trends*. Analog to what was said before, the trend rules in PHEN fire in parallel on the variables inserted during the variable consideration step. The strength parameters of the trend-rules are not changed; they have all the same strength value. After trends have been found (if any), control is returned to CONTROL once again, and it now triggers the application of function consideration rules by setting a respective goal and giving control to PHEN. Recall that the strength of function-consideration rules is modified according to hypotheses' success, so that only the currently strongest function-consideration rules will be selected and applied, again in parallel. Finally, the function-rules themselves are activated and insert new hypotheses into HMEM.

One additional point needs to be mentioned. When the goal *Generate Hypotheses* is set (that happens after a wrong hypothesis became discriminated, see Figure 6.11) the error-count is tested. If it's greater or equal 3.0[11], the strength parameter of all consideration rules will be set to 1.0, their initial value. That means, when the system has made three wrong predictions in sequence, it will again consider all possibilities for phenomenon generation. This limit lowers the probability that the system will get lost in unpromising areas of function induction.

So much for the basics of bucket brigading. HDD-SH*O(perator)P(reference)* is summarized in Table 6.7. Note that HDD-SHOP again uses all four function induction operators, since I want to test whether the bucket brigade algorithm restricts hypothesis generation effectively.

In a run with the factorial design, HDD-SHOP generated 84 hypotheses; 48 of them were general hypotheses, of which 7 became further discriminated, two of them twice. Table 6.8 shows the three hypotheses with the highest final strength value. Note that this model produced almost only correct final hypotheses in that two of the three correct hypotheses to be found in REFRACT are clearly elevated over the many other hypotheses found. This may be compared with the final hypotheses as produced by HDD-SH (cf. Table 6.5), where the final hypothesis set is much more undifferentiated. While the introduction of the bucket brigade mechanisms improved the quality of the final knowledge state, HDD-SHOP generates many spurious hypotheses. In order to understand why this model still produces so many irrelevant functional hypotheses we have to look at some of its process aspects.

---

[11] This is the current setting, I don't have a particular justification for it.

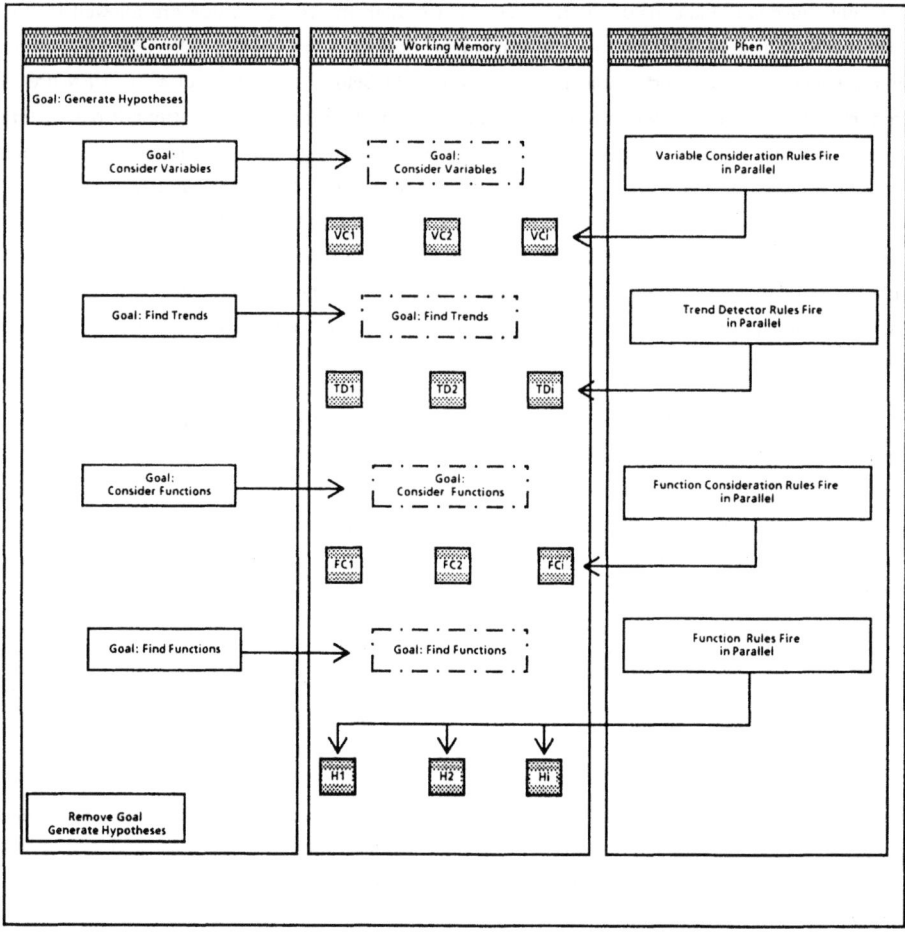

Fig. 6.12: Interaction of memories CONTROL and PHEN during the hypothesis generation step

Table 6.7

Architecture for learner model HDD-SHOP

PREDICTION MAKING	
Conflict Resolution Scheme	Select_All-Best(Strength)
	Select_One_Best(Specifity)
Forgetting in HMEM	Strength < .5
HYPOTHESIS MODIFICATION	
Learning Method	Discrimination
Limit on no. of conditions	LEQ 2
Strength Revision	
Increase	PLUS 1.0
Decrease	TIMES .25
BUCKET BRIGADE	
Initial Value	**1.0**
Increase	**PLUS 1.0**
Decrease	**TIMES .25**
Recovery Time	**3 wrong predictions in sequence**
HYPOTHESIS GENERATION	
Initial Strength	1.0
Increase When Rebuilt	1.0
Trend Detectors	Directly Proportional,
	Inversely Proportional
Mathematical Operators	PLUS, DIFFERENCE, TIMES, QUOTIENT

Table 6.8

Final hypotheses with strength value greater of equal 2.0

Hypothesis	Condition	Equation	Strength
H-37-4[a]	Subst. glass	Theta2=Theta1*-.4	2.0
H-51-2	ODist -200	Theta2=Theta1*-.33	2.0
H-51-4[a]	Subst flint	"	2.0

[a]Correct

An illustration how the bucket brigade mechanisms works is given in Figure 6.13, tracing HDD-SHOP through the first five experiments. In experiment exp-2, all variables and all functions are considered, leading to the generation of eight initial hypotheses. One of them, H-0031, is selected (on a random base) for prediction making, contributing to correct predictions in exp-3 and exp-4. Note that the consideration rules have their strength value increased together with the strengthening of H-0031. In exp-5, where surface form switches from plane to convex (Radius 100), H-0031 leads to a wrong prediction. Its strength value is lowered and so is the strength of its consideration rules. In the hypothesis generation step that follows the discrimination step, variables *Gamma* and *Alpha* as well as the function-operator *QUOTIENT* are not considered since their strength value is lower (.75) than the one of the other consideration rules (1.0). Hence, only two new hypotheses are built and one is rebuilt.

While HDD-SHOP's bucket brigading is quite useful in order to recover from following hypotheses with a wrong *phenomenon* description (as in the example above), it has some difficulties in accounting for errors due to an incorrect *condition* part. The problem is that in this case the phenomenon part is actually correct, but since the hypothesis is overly general, it has its strength decreased, as is the strength of the considerations rules. The system gets into a dead-end street with respect to function induction from which it recovers only by resetting all consideration rules after the next two wrong predictions.

In sum, the bucket brigading mechanism suggested here is a step in the right direction. In order to make it even more effective, the idea of focusing the system on particular elements of the problem space by distributing credit and blame must be generalized to consider the hypothesis *modification* process as well. That is, the bucket brigading mechanisms ought to be used not only to select variables and operators that enter the phenomenon part of hypothesis-rules, but also to select variables for the discrimination algorithm. I did not further elaborate this idea in my thesis project, but see it as a promising approach for further research. HDD-SHOP concludes my considerations about 'numerical' discovery models for REFRACT.

6.5 Summary: Quantitative Discovery Models

The computer models discussed so far all work within the same representation for experiments and hypotheses (see Table 6.1): Experiments are represented as sets of attribute-value pairs, where the attributes stand for variables from REFRACT, and where the values are mostly numbers. Hypotheses summarize experiments in form of simple equations relating a dependent variable to an independent variable, i.e., they perform function induction. Further, the production system architecture for all models is basically the same: There is a memory representing the information about experiments as it is stored in the *NoteBook*, distinguished from a second, more limited, working memory; and a distinction is made between knowledge used for discovery learning and knowledge acquired through discovery learning: separate production memories are defined for the two components.

Based on this representation format and core architecture, I introduced four computer models that induce numerical hypotheses resulting from the analysis of experiments and the comparison of prediction and feedback. The discrimination learning mechanism was kept constant over all models, while the function induction component was modified. Also invariant was the sequence of experiments used to illustrate the models' performance. We saw that all models can in principle find the regularities in REFRACT, given enough experiments. We saw further that differences in the way functional relations are induced lead to differences in discovery behavior and in the resulting knowledge structure (i.e. hypothesis-rules acquired). The first model, HDD-SH, used the function-induction algorithm as described in Chapter 5. Its main problem is that is has to

```
Experiment exp-1
[ 4]Design: glass plane v:-100.0 v:10.0 v:-10.0
[ 8]Feedback: v:4.0 v:4.0 v:-250.0 W

Experiment exp-2
[15]Design: glass plane v:-100.0 v:15.0 v:-15.0
[19]Feedback: v:6.0 v:6.0 v:-250.0 W
[34] Rule newly built:
Hypothesis-0031 (Hypothesis: DV: Gamma IV: Alpha Form: (LAMBDA (iv) (ROUNDED.QUOTIENT iv 2.5)))
[34] Rule newly built:
Hypothesis-0032 (Hypothesis: DV: Gamma IV: Alpha Form: (LAMBDA (IV) (ROUNDED.DIFFERENCE IV 9.0)))
[34] Rule newly built:
Hypothesis-0033 (Hypothesis: DV: Theta2 IV: Alpha Form: (LAMBDA (iv) (ROUNDED.QUOTIENT iv 2.5)))
[34] Rule newly built:
Hypothesis-0034 (Hypothesis: DV: Theta2 IV: Alpha Form: (LAMBDA (IV) (ROUNDED.DIFFERENCE IV 9.0)))
[34] Rule newly built:
Hypothesis-0035 (Hypothesis: DV: Gamma IV: Theta1 Form: (LAMBDA (iv) (ROUNDED.TIMES iv -.4)))
[34] Rule newly built:
Hypothesis-0036 (Hypothesis: DV: Gamma IV: Theta1 Form: (LAMBDA (IV) (ROUNDED.PLUS IV 21.0)))
[34] Rule newly built:
Hypothesis-0037 (Hypothesis: DV: Theta2 IV: Theta1 Form: (LAMBDA (iv) (ROUNDED.TIMES iv -.4)))
[34] Rule newly built:
Hypothesis-0038 (Hypothesis: DV: Theta2 IV: Theta1 Form: (LAMBDA (IV) (ROUNDED.PLUS IV 21.0)))

Experiment exp-3
[40]Design: glass plane v:-200.0 v:10.0 v:-10.0
[44]Feedback: v:4.0 v:4.0 v:-500.0 W
[48] (Prediction-0039 ISA: Prediction DV: Gamma IV: Alpha Form: (LAMBDA (iv) (ROUNDED.QUOTIENT iv 2.5)
) Value: 4.0)based on hypothesis Hypothesis-0031 is correct
[48] The new attribute value for the rule Hypothesis-0031 is: 2.0
[48] Attribute STRENGTHof TREND.ConsiderGamma changed: (2.0)
[48] Attribute STRENGTHof TREND.ConsiderAlpha changed: (2.0)
[48] Attribute STRENGTHof CONSIDER.Quotient changed: (2.0)

Experiment exp-4
[54]Design: glass plane v:-200.0 v:15.0 v:-15.0
[58]Feedback: v:6.0 v:6.0 v:-500.0 W
[62] (Prediction-0040 ISA: Prediction DV: Gamma IV: Alpha Form: (LAMBDA (iv) (ROUNDED.QUOTIENT iv 2.5)
) Value: 6.0)based on hypothesis Hypothesis-0031 is correct
[62] The new attribute value for the rule Hypothesis-0031 is: 3.0
[62] Attribute STRENGTHof TREND.ConsiderGamma changed: (3.0)
[62] Attribute STRENGTHof TREND.ConsiderAlpha changed: (3.0)
[62] Attribute STRENGTHof CONSIDER.Quotient changed: (3.0)

Experiment exp-5
[68]Design: glass v:100.0 v:-100.0 v:10.0 v:-20.5
[72]Feedback: v:8.2 v:-2.1 v:500.0 W
[76] (Prediction-0041 ISA: Prediction DV: Gamma IV: Alpha Form: (LAMBDA (iv) (ROUNDED.QUOTIENT iv 2.5)
) Value: 4.0)based on hypothesis Hypothesis-0031 is wrong
[76] The new attribute value for the rule Hypothesis-0031 is: .75
[76] Attribute STRENGTHof TREND.ConsiderGamma changed: (.75)
[76] Attribute STRENGTHof TREND.ConsiderAlpha changed: (.75)
[76] Attribute STRENGTHof CONSIDER.Quotient changed: (.75)
[76] Attribute SIMPLICITYof Hypothesis-0031-1 changed: (1.0 13 3.0)
[76] Discriminated rule: Hypothesis-0031-1
((GOAL: predict)
 (Exp =e Substance =iv1)
 (Exp =e Radius =iv2)
 (Exp =e ODist =iv3)
 (Exp =e Alpha =iv4)
 (Exp =e Theta1 -15.0))
[76] Discriminated rule: Hypothesis-0031-2
((GOAL: predict)
 (Exp =e Substance =iv1)
 (Exp =e Radius =iv2)
 (Exp =e ODist -200.0)
 (Exp =e Alpha =iv4)
 (Exp =e Theta1 =iv6))
[76] Attribute SIMPLICITYof Hypothesis-0031-3 changed: (1.0 13 3.0)
[76] Discriminated rule: Hypothesis-0031-3
((GOAL: predict)
 (Exp =e Substance =iv1)
 (Exp =e Radius plane)
 (Exp =e ODist =iv3)
 (Exp =e Alpha =iv4)
 (Exp =e Theta1 =iv6))
[91] Attribute SIMPLICITYof Hypothesis-0042 changed: (1.0 12 3.0)
[91] Rule newly built:
Hypothesis-0042 (Hypothesis: DV: IDist IV: ODist Form: (LAMBDA (IV) (ROUNDED.DIFFERENCE IV -600.0)))
[91] Rule: Hypothesis-0037 rebuilt.
[91] Attribute SIMPLICITYof Hypothesis-0044 changed: (1.0 12 2.0)
[91] Rule newly built:
Hypothesis-0044 (Hypothesis: DV: Theta2 IV: Theta1 Form: (LAMBDA (IV) (ROUNDED.PLUS IV 28.7)))
```

Fig. 6.13: Trace of HDD-SHOP from experiment 1 to experiment 5

consider very many hypotheses in the course of learning, mainly because it generates too many equations that capture spurious relations in the environment. I then described HDD-SHR and HDD-SHLR, versions that generated less hypotheses since they employed a restricted set of function-induction operators. Finally, HDD-SHOP was described, a model that was based on more general considerations about how to reduce the explosion of the number of hypotheses.

All these models work fairly 'syntactically', that is, besides information about experiments they use no further domain-related information. And they are not equipped with sophisticated data-analysis knowledge. These two characteristics lead to a at times unguided search for functional relations. In order to constrain the models even more and, hence, make them resemble the behavior of human subjects more closely, would require to either equip them with more knowledge about data-analysis, or to provide them with more background knowledge (or both). The first modification would move HDD into the direction of systems like BACON or ABACUS. But these systems are more appropriate to reflect the knowledge of trained discoverers, not that of untrained discoverers I'm interested in. The second modification, to add background knowledge, would require to add certain biases in form of preferences to the model, preferences for the form of equations, operators and variables that are not acquired in the course of learning in REFRACT, but that the discovery learner brings to the task based on former experience. That is, it amounts to take into account the learning-history of the learner before he enters the specific learning situation I focus on. In my current framework, I do not have the means to consider this kind of learning background, so the only way to go for me would be to introduce such preferences ad hoc. I decided not to do that, but concentrate instead on an extension of the framework in another direction: the generation of qualitative hypotheses. Before I turn to this question, let me mention a final quantitative model.

A Model That Derives Multiple Predictions: HDD-MH

The last numeric model to be discussed is not meant to be a plausible cognitive model of the learning task in REFRACT. Rather, it it described here to demonstrate a beam-search strategy trough the hypothesis space. The model variant tests *multiple hypotheses* in parallel (-MH stands for *Multiple Hypotheses*) by deriving a set of predictions instead of a single one and by comparing this set of predictions with the experiment outcome. From what is known about human concept formation and scientific discovery, testing multiple hypotheses is a rarely used strategy (Bruner et al., 1956; Mynatt et al., 1978).

Implementing HDD-MH serves also as a demonstration that the HDD implementations are written in a fairly general manner that supports rapid construction of model variants. The only thing that needs to be changed to switch from best-first search (as in HDD-SH and HDD-SHOP) to a breadth-first search for hypotheses is the conflict-resolution scheme for HMEM. Specifically, the conflict-resolution has to be defined so that not a single best hypotheses is selected, but all the hypothesis-rules that tie on the ordering dimension (first strength, then specifity). In other words, in HDD-MH all those hypothesis-rules will fire that have the maximal strength value and then the maximal specifity value.

Not surprisingly, the number of hypotheses considered by HDD-MH grows quickly. For example, in a run with the factorial design, overall 101 hypotheses were generated in 17 experiments: 64 unconditional hypotheses, of which 9 were further discriminated, one of them two times. All those nodes (here: hypothesis-rules) in the search space that score best on the evaluation function 'strength' became expanded. This kind of search does quickly lead to building the correct hypotheses, but it requires a lot of working memory space. Limitation of working memory is the argument mostly put forward to explain why humans do not seem to consider multiple hypotheses at a time (e.g., Mynatt et al., 1978).

6.6 Generating Qualitative Hypotheses

Differences in problem representation in REFRACT concern to a substantial degree differences in the way subjects perceive the entities in the discovery world. As we have seen in analyzing two subjects' protocols and by looking for evidence in protocols of other subjects, a prominent difference was whether numeric information was included in their representations of experiments. This difference can be expressed as a bias on the observation language. Different subjects use different descriptors to encode what they perceive. In more psychological terms, subjects may have different conceptual knowledge that determines the kind of information they can encode. On the following pages, I consider the implications of such differences in terms of process characteristics and learning outcome. Note that I go with this analysis beyond the framework that was introduced in the task analysis in Chapter 5.

6.6.1 Preference For Pictorial Descriptions

For the domain of REFRACT, environmental states correspond to the information on the screen about ray-experiment designs, predictions, and experiment outcomes. I call the rules that map from the design and outcome descriptions into categories of the learner's mind *encoding rules* and the rules that allow the subject to predict the path of rays *hypotheses*.

It is important to note that the mapping from the environmental states to the categories in the mental model establish a form of information selection. Under almost all circumstances, only a small part of the environmental information is represented in the mental model. But this is the only information available to the learner to come up with predictive rules. Hypotheses can only be built based on what is encoded and focused on at any given time. We may think of the encoding rules as providing a vocabulary to describe states of the world. The rules used by a learner to represent the domain establish a specific view of the environment: a descriptive bias.

6.6.2 A Pictorial Description Language

To make the discussion more concrete, I want to develop an alternative representation language for screen information. This will be done by modifying the notation introduced in Chapter 5 so that it becomes less numerical, while preserving the attribute-based format of the notation. In general, my intention is to develop a - although primitive - pictorial representation for the screen objects, where pictorial means a representation that preserves the spatial topology of the problem so that "each component occupies a specific position and holds specific spatial relations with other components" (Langley et al., 1987, p. 321; see also Larkin & Simon, 1987).

The numerical notation was summarized in Table 6.1. An alternative description language is shown in Table 6.9. Most of the attributes have now a reduced value set, often only two or three values. The idea is that encoding[12] of screen information based on this description language will be done mainly in terms of approximative distance estimations and ray directions instead of precise values for angles and distances. Thus, a learner might perceive the refracted ray as going *upwards* instead of 'seeing' that $Theta2 = 12.3$ *degrees*.

Table 6.9

The 'pictorial' description language

Attribute	Value-set
Substance	(Glass, Flint, Diamond)
Shape	(plane, convex, concave)
ODist	(Close, Far)
Alpha	(Up, Down)
Theta1	(Up, Straight, Down)
Theta2$_{E/H/N}$	(Up, Straight, Down)
Gamma	(Up, Straight, Down)
IDist	(Close, Far)

For the descriptor that represents the angle of refraction (*Theta2*), an additional variation is possible: The reference line the angle is measured against can be varied. This is indicated in Table 6.9 by the indices *E*, *H*, and *N*. Let us assume that graphical information is encoded quite simply in form of the direction of the refracted ray. The subject may describe rays as running *straight*, going *downwards*, or going *upwards* relative to a reference line. Reference lines for the refracted ray may either be the *extension* of the incoming ray (*E*), the *horizontal line* through the point of intersection between the incoming ray and the medium (*H*), or the *Normal* line (*N*). Thus, the path

---

[12] I do not claim that what is introduced here is a theory of spatial perception or spatial reasoning. My assumptions about how subjects encode the objects rest on intuitions nurtured from protocol reading rather than on a well-founded body of empirical evidence. Similar to Larkin & Simon (1987), I would only claim that the elements in the 'graphical' representation are "perceptual obvious to any educated person" looking at the screen.

of the refracted ray can be described in three different ways, or with three different attributes[13]. Figure 6.14 depicts the three reference lines.

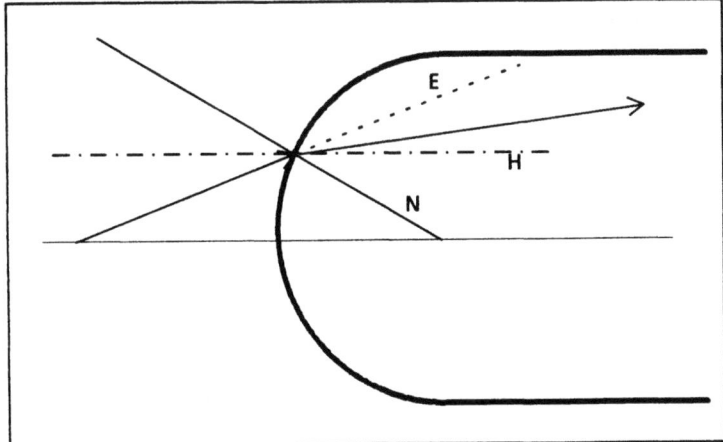

Fig. 6.14: Three different reference lines to describe the direction of the refracted ray

Other 'pictorial' description languages are conceivable. For example, distance might be represented as (far, middle, near). But the representation developed above should suffice as an example. Given the possibility of different languages for encoding experimental data, we can speculate about the influence their usage will have on learning and on the resulting knowledge structure.

6.6.3 Effects of Representation on Inductive Problem Solving

I want now address the question: What effects do differences in representation have on learning? The answer will be that both the learning outcome, i.e., the content of final hypotheses, and the process of learning in the form of hypothesis generation and testing is affected. For illustration purposes, the two description languages shown in Tables 6.1 ('numerical') and 6.9 ('pictorial') will be compared on a number of dimensions.

Effects on the Learning Process

Effects with respect to both the final results and the learning process stem from the fact that the graphical and numerical representation format are not informationally equivalent, i. e., it is not the case that "all of the information in the one [representation] is also inferable from the other" (Larkin & Simon, 1987, p. 67). The two forms of representation are not equivalent in a second sense: they are not computationally equivalent. Larkin and Simon define computational

---

[13] All three forms were observed in subjects' statements

equivalence as follows:

> Two representations are computationally equivalent if they are informationally equivalent and, in addition, any inference that can be drawn easily and quickly from the information given explicitly in the one can also be drawn easily and quickly from the information given explicitly in the other, and vice versa. (ibid., p. 67).

Since the graphical and numerical representation format are not informationally equivalent, they cannot be, by definition, computationally equivalent. But even if they would be informationally equivalent we still would expect to see differences with respect to processing parameters. For one, it is easier within a pictorial description to derive a phenomenon description: it can be 'read from the screen'. Within a numerical description, it has to be calculated. However, parsimonious generalizations need to be based on numeric data, since this is the level where the most general regularities emerge. In other words, there is a trade-off between finding a phenomenon description quickly and easily, and finding parsimonious generalizations. The later need more effort to get created, but that pays off in that they are more generally applicable.

Effects on the Learning Outcome: Parsimony

Turning to the question of effects on learning outcome, recall that in the course of protocol analysis the final hypotheses of two subjects were compared with respect to the parsimony of knowledge representation using Holland et al's (1986) concept of a default hierarchy. The idea was to organize a knowledge structure into default rules and rules that cover exceptions. This concept is used here to speculate about the effects of a numerical versus a pictorial representation on the knowledge structure that can be acquired. I define a knowledge structure $K1$ as being more parsimonious than another structure $K2$ if $K1$ consists of less rules than $K2$ and if $K1$ allows to generate the same predictions as $K2$, or more.

The description language used imposes a lower limit on the number of rules (my smallest unit of knowledge representation) that have to be generated to account for all observations in the microworld. We do know how many rules are needed to predict rays based on numerical features: one - the law of refraction - in the ideal case; three or four for more realistic cases (cf. Subject S10's final rules in Table 4.12). I want now to estimate the minimal number of rules one needs in order to predict all paths of the refracted ray based on the pictorial description language from Table 6.9.

I counted how often a refracted ray changes its direction from *straight* to *upwards*, *upwards* to *downwards*, and so on, for all incoming rays with positive *Alpha*, and this for all three reference lines. Table 6.10 shows the result for *flintglass*. According to this table, four rules are needed to cover the changes in ray direction for the reference system *Extension*, five rules for the reference system *Horizontal*, and four rules for *Normal*. This example covers only a single medium type (*flintglass*) and only half of the number of possible input rays (only positive ones). The number of rules necessary to describe the pictorial information would grow if, for example, subjects notice the slight difference of 'bending' in different substances. In other words, mental models of the domain based on pictorial information will include much more rules to cover observations than will models based on a numerical description language.

Table 6.10

Ray direction changes for three reference systems

Surface	ODist	Alpha	Extension	Horizontal	Normal
Plane	50	10	D	U	U
		15	D	U	U
		20	D	U	U
	100	10	D	U	U
		15	D	U	U
		20	D	U	U
	200	10	D	U	U
		15	D	U	U
		20	D	U	U
Convex	50	10	D	S	U
		15	D	S	U
		20	D	U	U
	100	10	D	D	U
		15	D	D	U
		20	D	D	U
	200	10	D	D	U
		15	D	D	U
		20	D	D	U
Concave	50	10	D	U	U
		15	D	U	U
		20	D	U	U
	100	10	S	U	U
		15	S	U	S
		20	S	U	S
	200	10	U	U	D
		15	U	U	D
		20	S	U	S

Note. Medium is always flint glass. The three reference systems are: Extension of the incoming ray, Horizontal through the point of intersection, and Normal line. The direction of the refracted ray is coded as going Straight, Up, or Down.

6.6.4 A Qualitative Version of HDD: HDD-QUAL

Building on these considerations about a pictorial representation language, I wrote a production system simulating a discoverer who represents experiments and hypotheses in terms of the pictorial language outlined in Table 6.9. In particular, this artificial learner - coined HDD-*QUAL(itative)* - represents substance as (glass, flintglass, diamond), radius as (plane, convex, concave), object distance as (close, far), and *Alpha* as (up, down). The direction of the refracted ray is described as (above-horizontal, on-horizontal, below-horizontal), i.e., the *Horizontal* is used as the reference system (Figure 6.15). This preference for the *Horizontal* instead of for the *Normal* was found in children of various age groups in other studies with the optics domain (Spada, Reimann, & Häusler, 1983).

HDD-QUAL is based on the same architecture as the HDD-SH version. Knowledge is distributed over the four memories CONTROL, HMEM, WM and ENV. The representation of hypotheses and predictions needs to be modified. A prediction has now the form:

        (Prediction-<number> ISA: Prediction FORM: <ray path>),

where <ray path> is one of *above-horizontal, below-horizontal,* or *on-horizontal.* Hypothesis-rules contain a corresponding expression on their right-hand side which they insert into Working Memory when they fire.

The productions that work under the goals DESIGN-EXPERIMENT, PREDICT, COMPARE-PREDICTION-FEEDBACK, and MODIFY-HYPOTHESIS are only slightly modified to account for the new representation format. The many productions needed to generate a hypothesis based on a numeric format are replaced by a single production:

```
IF      the goal is to generate hypotheses,
        and the current experiment is =e,
        and in =e the refracted ray has direction =d
THEN    build a new rule with a general condition part and a phenomenon
        part saying that the refracted ray will take direction =d
```

This rule creates a new hypothesis by 'copying' the current ray path into a phenomenon description. Consequently, the resulting production system is much shorter than the other HDD versions; trend detection and function induction rules are removed, as well as the rules required to manage the flow of control between memories PHEN and CONTROL. The production system is listed in Appendix VI.4.

In order to experiment with HDD-QUAL, I had it run with a factorial design comprising 16 experiments based on the factorial levels (1) *Alpha* (up, down), (2) *ODist* (close, far), (3) *Radius* (convex, concave), and (4) Substance (glass, diamond). In the course of running these 16 experiments, the system acquired 14 hypotheses, listed in Table 6.11.

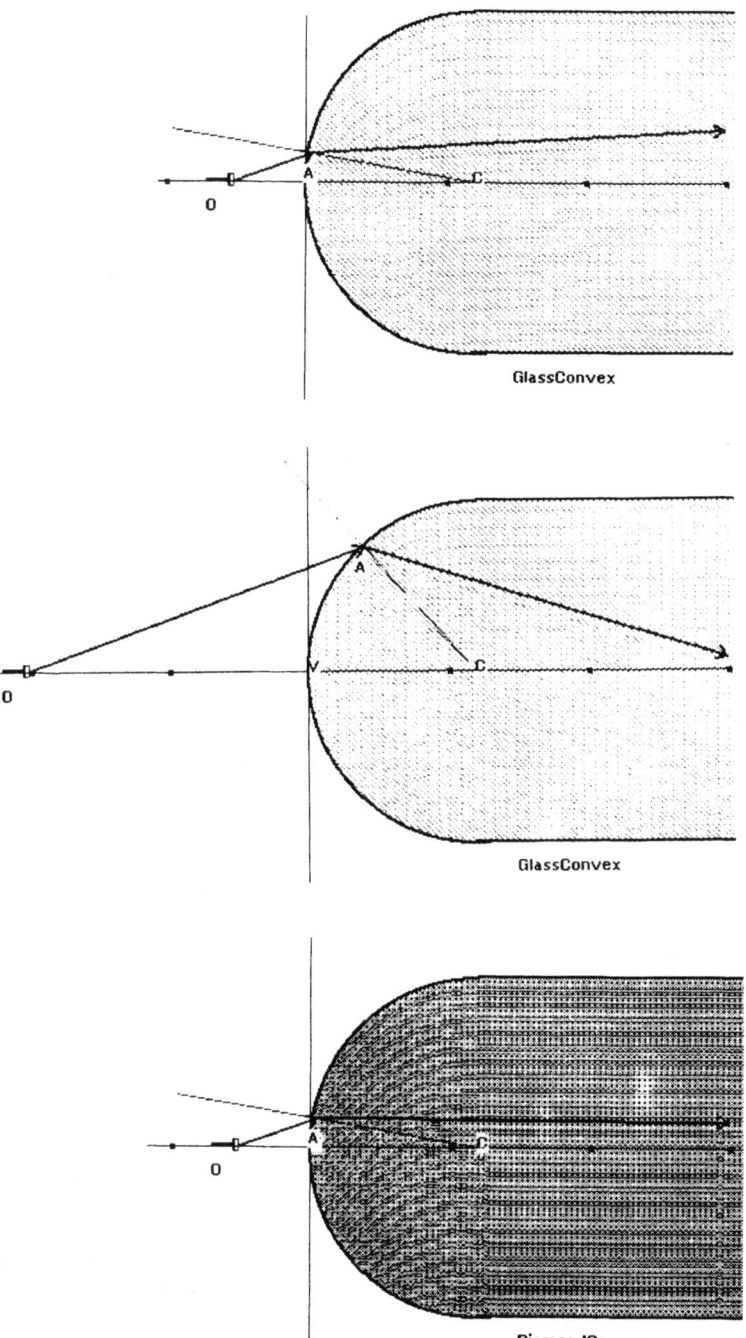

Fig. 6.15: Example for ray paths *above-horizontal*, *below-horizontal*, and *on-horizontal*.

Table 6.11

All final hypotheses of HDD-QUAL

Hypothesis	Condition	Ray Path	Strength
H-62		above-horizontal	.5
H-62-1	Alpha down	"	1.0
H-62-2	ODist close	"	.3
H-62-2-1	ODist close & Radius concave	"	4.0
H-62-2-2	ODist close & Subst. glass	"	4.0
H-65		below-horizontal	.5
H-65-1	Alpha down	"	.5
H-65-1-1	Alpha down & Radius convex	"	1.0
H-65-2	ODist far	"	.25
H-65-3	Radius convex	"	1.0
H74		on-horizontal	.5
H74-1	Alpha down	"	1.0
H74-2	ODist close	"	1.0

As can be seen from Table 6.11, rules with a high strength parameter all contain two restricting conditions (no limit was set on the number of conditions a hypothesis can have). This is so because the pictorial encoding does not allow to describe the regularities more parsimoniously. This is the crucial difference to the numeric presentation, where a single condition specifying substance suffices to predict the ray path correctly and precisely.

6.7 Why Do Learners Prefer Pictorial Representations?

This needs to be asked because subjects can access numerical information in REFRACT easily and with no costs. However, most of them prefer at least temporarily the graphical information. Two explanations are possible here. The first considers top-down processing. It may very well be the case that a learner does not have the knowledge available that is necessary to analyze numerical data. For example, he might not know how to find a linear trend in data, or he might not activate that knowledge. Under these circumstances it makes sense to prefer pictorial information over numerical.

A second explanation considers bottom-up processing, building on the notion of perceptual salience. In REFRACT, pictorial information is simply more salient to the learner. Without knowing anything about the labels the system uses (*Alpha, Theta1*, etc.) one can readily understand what happens on the screen: The ray 'gets bent' in the one or the other form. This notion is closely related to what Larkin (1983) calls the 'basic representation' of a problem, i.e., "the collection of objects and operators suggested directly by the problem statement", or, in our case, by the screen. We can think of subjects who utilize numerical information as activating an additional set of rules besides those that allow them to encode the pictorial information. This additional rule set enables them to elaborate the picture and to 'go beyond the information given'. Research on problem solving (Simon & Hayes, 1976; Reed, Ernst & Banerij, 1974; Larkin, 1983) shows that "people do not find it easy to translate between representations and therefore usually work with the basic representation directly suggested by the problem statement" (Larkin, 1983, p. 2). Hence, differences between learners with respect to their attempts to elaborate the basic representation might lead to the differences in description languages employed.

The notion of 'salience' of information can be operationalized more concretely by modeling in detail what students in REFRACT see on the screen at any given point in time. For example, recall that the *NoteBook* displays only six consecutive experiments and only information about variables actively entered by the student. Hence, the information configuration might often be of a kind so that trend-detection and function-induction operators can not get triggered because their condition part (which describes a pattern of experimental data in the *NoteBook*) is not satisfied. Other peculiarities of the domain presentation are also psychologically relevant. For instance, the circumstance that only one experiment can be seen graphically on the screen certainly puts limitations on what can be learned within a pictorial representation format, in particular when considering students' memory limitations. Considerations of this kind lead us to fascinating questions regarding the interplay between the way information is laid out on the screen, students' attention processes and memory limitations, and students' knowledge about experimenting and data analysis. Though I did not have a chance to follow these questions into more detail in the context of my thesis research, further research into these issues is possible based on the structure of the simulation models outlined in this chapter (for example, by relaxing the unrealistic constraint that experiments are always represented by 'rows')

6.8 Changes in Problem Representation and the Case of Semi-Quantitative Hypotheses

So far I described a quantitative and a qualitative ('pictorial') format to represent experiments and hypotheses and speculated about why it might be that learners stick to a qualitative representation. I did not say *why* and *how* learners might move from a qualitative to a quantitative representation format, and vice versa.

The question of why a learner might switch between the two formats is more easily answered. For the first direction, from qualitative to quantitative: Precise predictions can only be made based on a quantitative representation. In the study with REFRACT, both the system (by means of its feedback distinction *approximately* correct versus *correct*) and the experimenter urged the student to improve prediction precision over time. Further, besides this external pressure there might be an learner-inherent, knowledge related reason to move to quantitative predictions: As my analysis has shown, in order to predict the ray path pictorially, one has to store many exceptional cases and the student may at some point feel the need to look out for a more parsimonious description of regularities.

The step from a quantitative representation to a qualitative representation seems mostly to occur when a student has designed a experiment for which he knows that his current knowledge is not sufficient to derive a precise prediction. See Subject S01's and S10's behavior as described in Chapter 4.

The second question, how learners manage to come to a quantitative representation the first time is much harder to answer. I reckon that they build an intermediate representation, a semi-quantitative one where the main entity is covariation information: *When x decreases, y increases*. Such assertions might even be derived from pictorial information: *When the object distance increases, the ray is less bent*, and so on. In order to represent such hypotheses built on change-information, we would have to extend HDD's use of trend-detectors. Currently, trend-detectors are only used to prepare the ground for function induction (see the description of HDD-SH, for example). In order to make them directly useful for hypothesis generation and prediction derivation, the representation language for a hypothesis' condition part must be modified. The condition side must describe value-changes between experiments, for example:

```
IF      between experiment =e1 and =e2 the value of independent variable =i
        increases,
        and ... (further conditions)...
THEN    predict that the value of dependent variable =d will increase from
        =e1 to =e2.
```

However, these are again suggestions for further research, suggestions I cannot follow in this thesis.

6.9 Conclusions

I showed how the conceptual considerations from Chapter 5 can be translated into executable computer programs; how processes of hypotheses generation and modification in REFRACT can be realized as production system programs (processes of experiment generation were not treated). This was done by expressing assumptions about the cognitive architecture of a discovery learner in form of a production system architecture, expressing assumptions about problem representation in terms of a specific description language for experiments, representing knowledge about how to work in REFRACT in form of production rules that manipulate descriptions of the environment and internally considered goals, and by implementing learning mechanisms in form of productions rules and Lisp code.

Besides providing programs that accomplish the task, (i.e., find Snell's Law as present in REFRACT), I began to analyze the effects different approaches to the discovery problem have both on the process and the outcome of learning. This was done computationally by developing different versions of computerized learners and comparing their discovery behavior and their resulting hypotheses in the context of an experiment sequence which was identical for all models. The models discussed differed along two main dimensions: quantitative versus qualitative representation of the objects in REFARCT, and, within the quantitative representation: constraints on the function induction step. The comparison of the quantitative with the qualitative ('pictorial') approach to induction in REFRACT revealed the same finding that was made in the context of the empirical study: A pictorial representation requires more and more specific hypotheses in order to lead to (though only approximately) correct predictions than the numerical representation. The comparison of numerical models that employed different constraints to induce functional relations revealed trade-offs between the absolute number of rules generated and degree of specifity of the most successful rules. In order to explain how certain biases (or, more positively stated, preferences) towards function induction may be acquired in the course of discovery learning in REFRACT, I introduced a computer model that encompasses a method to distribute credit and blame over the function construction rules (see the summary in Section 6.5).

7. General Discussion

Summary of the Thesis

I analyzed a discovery learning task that includes components relevant for scientific discovery and induction in general. It took the form of a graphical simulation program for refraction phenomena: REFRACT. My investigation started from empirical observations gathered in an exploratory study with REFRACT. Subjects in this small study showed many differences, both with respect to knowledge acquired and to learning behavior. On side of the learning process, I noticed differences with respect to control of variables, prediction making, discrimination of relevance, and determination of covariations. As for the knowledge acquired, most striking was that many subjects were not able or not willing to formulate hypotheses which are precise in the sense that they are based on numerical relationships between variables; instead, they expressed hypotheses 'pictorially' or semi-quantitative. I made the lack of knowledge about how to analyze numerical data to one of main the features in the computational models I developed for the task. Further, I decided to treat the issue of 'pictorial' hypotheses in more detail.

The observations on human discovery learners were formulated as constraints on a conceptual information processing model of the task, the *Hypothesis Driven Discoverer*, which was in turn used to construct computational models of discovery learning in REFRACT. These production system models of discovery learning were based on a common core architecture. I introduced four computer models that induce numerical hypotheses resulting from the analysis of experiments and from the comparison of prediction and feedback, and that correct for overgeneralizations by means of discrimination learning. All these models work fairly 'syntactically', that is, besides information about experiments they use no further domain-related information; also, they are not equipped with sophisticated data-analysis knowledge. Thus, they stand for the untrained discovery learner that has no knowledge about optics.

All of these models were able to discover the regularities in REFRACT. But even on the abstract level on which I compared human with computer discovery learners, it was obvious that the human learners attacked the problem in a more constrained and biased way than did the computer models. I attempted to reconstruct these biases on induction of functional relations by specifying learning processes - besides hypothesis acquisition - that can take place while working with REFRACT. From observations of how this more advanced model learned I drew the conclusion that the biases my subjects employed must probably be traced back to pre-experimental knowledge that they bring to bear on the discovery learning task. I described these biases as preferences for certain ways to induce functional relations, preferences which were not developed while learning in REFRACT, but are based on experiences in former learning situations. I did not further elaborate on the sources for such preferences, but showed in an exemplary manner the implications they have for inductive knowledge acquisition in my discovery task. Besides showing how preferences for certain ways to capture quantitative relations in data influence discovery learning and its outcome, I also illustrated the implications of preferring a qualitative, 'pictorial' style of stating hypotheses in REFRACT.

The Merits of 'Computational Experiments'

The main merit the computational approach to discovery learning brought to my analysis was that they enabled me to test fairly quickly whether certain assumptions about cognitive processing are *sufficient* to solve the task at hand. A second advantage is that they allowed to evaluate the plausibility of certain assumptions in the light of what we know about human induction. For example, I tried to show in Chapter 6 that formulating constraints on (function) induction in terms of filters on hypotheses leads to processing demands which require to consider more hypotheses than it seems possible for human learners, given their memory limitations. Finally, the computational style for developing theories of discovery learning provided me with means to speak more clearly about such issues as 'knowledge' (treated as (production rules)), 'representation' (I formulated representations of the task as problem spaces), and 'capacity limitations' (operationalized in my models as limits on production system memories).

As a side product of my cognitive modeling work, I developed some tools that support 'rapid prototyping' of different versions of a production system model, combining some ideas from frame-based information storage with the PRISM shell (Reimann, 1988d). It builds in turn on a menu-interface for PRISM that I developed (Reimann, 1988c).

Contributions

It is obvious that the kind of discovery learning addressed in this thesis covers only parts of what belongs to scientific thinking and induction in general. From the central activities comprised under 'scientific discovery'- gathering data, finding descriptive generalizations, formulating explanatory theories, and testing the theories--only two are treated in more detail: gathering data by experimentation, and finding descriptive (as well as predictive) generalizations, and only the later aspect was realized computationally.

However, the formulation of processing constraints as I attempted them in my modeling study holds the possibility that the analysis of this specific discovery problem may contribute to a better understanding of discovery and induction in general. Further, I tried to distinguish explicitly between domain specific requirements (e.g., in form of a specific representation language) and between what is assumed to be task-independent (e.g., assumptions about memories, about focus of attention, about general problem-solving and learning strategies, in general: about processing constraints). Finally, going from the conceptual model of the task to specific implementations should make it easier to judge what has to count as mainly technically motivated specifications and what are the more principled model assumptions.

This work might contribute to instructional research in two ways. For one, with REFRACT I have developed a simulated discovery environment that can be used (and has been used) by students to learn about optics and to learn about certain scientific reasoning skills. Seen as an instructional device, REFRACT could be improved in many directions. For example, playing back information to the student is currently realized as a *replay* that frees students from the singularity and time constraints of real-world events. More abstract forms of presenting problem solving behavior are conceivable such as *abstract replay*-- thereby controlling students' focus of attention--and *spatial reification,* where in addition to the simulation theoretical concepts and multiple representations are put to use (cf. Collins & Brown, 1986). For example, systems have been developed in algebra and geometry, where they provide a structured "trace" of problem solving behavior so that students can see the alternative paths they have tried (Anderson, Boyle, Farrell & Reiser, 1984; Brown, 1983). In Smith's *Alternate Reality Kit,* theoretical notions of physics such as force and gravity are reified as objects which can be inspected and modified by the student (Smith, 1986). For future versions of REFRACT, I plan to include facilities that allow for abstracted replay and reification of the inductive learning process. A proposition to be evaluated then is that effective interrogative skills are teachable if the particular skills involved can be articulated an practiced under circumstances which require them to be used.

A second contribution to pedagogical research is made in form of the detailed analysis of cognitive demands involved in inductive learning from simulations. Such a task analysis is directly relevant for the development of intelligent (computerized) tutoring and coaching system (ITS). Intelligent tutorial and training systems have to monitor the student's progress in order to provide him with optimal information. In the last decade, a number of systems have been developed that can form a model of the student based on dialogues and observations about his activities (cf. Reimann, 1989). Student modeling is often considered to be a special type of diagnosis, the goal of which is the construction of a *computational* model of students' knowledge (including their misconceptions) and - ideally - of changes in the knowledge states due to learning processes.

The research done in this thesis can be seen as a small contribution to developing a generative theory of how misconceptions in the domain of refraction arise. As generative mechanism that can give rise to misconceptions I identified: inadequate problem representation, insufficient knowledge about how to find regularities in data and how to gather data effectively, and learning factors (overgeneralization, overspecialization). These different factors were integrated into an overall model of human cognitive architecture, as it was developed in problem solving theory

Besides theoretical contributions to research on ITS, more practical consequences are conceivable. Since my models of discovery are implemented as computer programs, they can be made more directly useful for intelligent computer-provided instruction by serving as the basis of the diagnostic component of an ITS, i.e., the component that constructs automatically a model of the student's knowledge and skill. For example, starting from certain assumptions of the kind of problem space the student will work in, the computer can co-learn with the student, i.e., receive the experiments the student runs as input and have its inductive inference operators run over it. The output of the learning model would be - for every experiment - a hypothesis about the student's knowledge state in form of a set of currently dominating hypothesis-rules. This hypothesis about the student's hypotheses can be tested directly for each experiment by having the diagnosing component deliver a prediction and comparing this prediction with the student's prediction. The quality of the diagnosis can be estimated quickly by the hit/miss rate.

Such a student model could be put to use in the following scenario: A computerized coach might be build on top of the simulation world REFRACT. Such a coach (Burton & Brown, 1982) could serve as a critic on the students discovery behavior. For example, he could comment on the quality of the design suggested by the student, and evaluate a students predictions and hypotheses. An intelligent evaluation of an hypothesis, for instance, would require to check whether it conforms with all the observations made up to the current experiment. Further, if asked by the student the coach could suggest experiments on his own, based on an intelligent estimation of what it kind of information the student needs in order to progress optimally. All these coaching activities - there can be others and more - depend on having a model of the student's knowledge available in order to be optimally useful. For example, criticizing a student's hypothesis requires to see it in light of his current knowledge state. Evaluating or suggesting an experimental design to the student requires to see it in light fo his current knowledge state.

I do not believe that the current HDD simulation models would deliver diagnostic models of individual students with an acceptable quality. Too many factors that influence performance in REFRACT were left out in building the concrete implementations. However, the conceptual framework may proof valuable and additional knowledge sources may be integrated easily in order to accomplish the diagnostic task. This is a direction further research will take.

References

Anderson, J.R. (1976). Language, memory, and thought. Hillsdale, NJ: Erlbaum.

Anderson, J.R. (1983). The architecture of cognition. Cambridge, MA: Harvard University Press.

Anderson, J.R., Boyle, C., Farrell, R.& Reiser, B. (1984). Cognitive principles in the design of computer tutors. Pittsburgh, PA: Carnegie Mellon University.

Anderson, J.R.& Kline, P.J. (1979). A learning system and psychological implications. In: Proceedings of the Sixth International Conference on Artificial Intelligence.

Anderson, J.R., Kline, P.J., & Beasley, C.M., (1980). Complex learning processes. In R.E. Snow, P.A. Federico, & W.E. Montague. (Eds.), Aptitude, learning, and instruction: cognitive process analysis. Hillsdale,NJ: Lawrence Erlbaum.

Bower, G.H., & Trabasso, T.R., (1964). Concept identification. In J.R. Atkinson. (Ed.), Studies in mathematical psychology. Stanford, CA.: Stanford University Press.

Brown, J.S, (1983). Learning by doing revisited for electronic learning environments. In M.A. White. (Ed.), The future of electronic learning. Hillsdale, NY: Erlbaum.

Bruner, J.S., Goodnow, J.J.& Austin, G.A. (1956). A study of thinking. New York: Wiley.

Burton, R.R., & Brown, J.S., (1982). An investigation of computer coaching for informal learning activities. In D. Sleeman, & J.S. Brown. (Eds.), Intelligent tutoring systems. New York: Academic Press.

Carbonell, J.G., (1986). Derivational analogy: a theory of reconstructive problem solving and expertise acquisition. In R.S. Michalski, J.G. Carbonell, & T.M. Mitchell. (Eds.), Machine Learning. An artificial intelligence approach (Vol. 2). Los Altos, CA: Kaufmann.

Carey, S. (1984). Cognitive development. The descriptive problem. In Gezzaniga (Ed.), Handbook of cognitive neuroscience. New York: Plenum Press.

Chomsky, N. (1965). Aspects of the theory of syntax. Cambridge, MA: MIT Press.

Clancey, W.J. (1986). Qualitative student models. In J.F. Traub (Ed.), Annual Reviews of Computer Science (Vol. 1). . Palo Alto, CA: Annual Reviews.

Collins, A., & Brown, J.S., (1988). The computer as a tool for learning through reflection. In H. Mandl & A. Lesgold (Eds.), Learning issues for intelligent tutoring systems. New York: Springer.

Dietterich, T.G., (1982). Learning and inductive inference (Chapter XIV). In P.R. Cohen, & E.A Feigenbaum. (Eds.), The handbook of artificial intelligence (Vol. 3). Los Altos, CA: Kaufmann.

Falkenhainer, B.C, & Michalski, R.S. (1986). Integrating quantitative and qualitative discovery: the ABACUS system. Machine Learning, 1, 367-402.

Feigenbaum, E.A., (1963). The simulation of verbal learning behavior. In E.A. Feigenbaum, & J. Feldman. (Eds.), Computers and thought. New York: McGraw-Hill.

Gentner, D. (1983). Structure-mapping: A theoretical framework for analogy. Cognitive Science, 7, 155-170.

Gick, M.L., & Holyoak, K. (1983). Schema induction and analogical transfer. Cognitive Psychology, 15, 1-38.

Glaser, R., Bonar, J., Shute, V.& Lesgold, A. (1985). Exploration and tutoring in simulated science laboratories (Technical Proposal). Pittsburgh, PA: Learning Research and Development Center, University of Pittsburgh.

Gorman, M.E. (1984). A comparison of disconfirmatory, confirmatory, and control strategies on Wason's 2-4-6 task. Quarterly Journal of Experimental Psychology, 36A, 629-648.

Greeno, J.G.& Simon, H.A. (1984). Problem solving and reasoning (Technical Report). Pittsburgh: Learning Research and Development Center, Unversity of Pittsburgh.

Gregg, L., & Simon, H.A. (1967). Process models and stochastic theories of simple concept formation. Journal of Mathematical Psychology, 4, 246-276.

Halliday, D.& Resnick, R. (1986). Fundamentals of physics. New York: Wiley.

Hobbs, J.R.& Moore, R.C.(Eds.) (1985). Formal theories of the commonsense world. Norwood, NJ: Ablex.

Holland, J., Holyoak, K., Nisbett, R.E.& Thagard, P. (1986). Induction: Processes of inference, learning, and discovery. Cambridge,MA: MIT Press.

Huesman, L.R., & Cheng, C.M. (1973). A theory for the induction of mathematical functions. Psychological Review, 80, 126-138.

Hunt, E.V., Martin, J.& Stone, P.I. (1966). Experiments in induction. New York: Academic Press.

Johnson-Laird, P.N., Legrenzi, P., & Legrenzi, S. (1972). Reasoning and a sense of reality. British Journal of Psychology, 63, 395-400.

Keil, F.C. (1981). Constraints on knowledge and cognitive development. Psychological Review, 88, 197-227.

Kintsch, W., Miller, J., & Polson, P., (Eds.) (1984). Methods and tactics in cognitive science. Hillsdale, NJ: Erlbaum.

Klahr, D., & Dunbar, K. (1988). Dual space search during scientific reasoning. Cognitive Science, 12, 1-45.

Klayman, J., & Ha, Y. (1987). Confirmation, disconfirmation and information processing in hypothesis testing. Psychological Review, 94, 211-228.

Kotovsky, K., & Simon, H.A. (1973). Empirical tests of a theory of human acquisition of concepts for sequential patterns. Cognitive Science, 4, 399-424.

Kulkarni, D., & Simon, H.A. (1988). The processes of scientific discovery: the strategy of experimentation. Cognitive Science, 12, 139-175.

Langley, P., (no year) The dimensions of learning: An analysis of condition-finding methods (Technical Report). Pittsburgh,PA: The Robotics Institute, Carnegie-Mellon University.

Langley, P. (1979a). Descriptive discovery processes: experiments in Baconian science (Doctoral Dissertation). Pittsburgh,PA: Carnegie-Mellon University.

Langley, P. (1979b). A production system model for the induction of mathematical functions. Behavioral Science, 24, 121-139.

Langley, P., (1987). A general theory of discrimination learning. In D. Klahr, P. Langley, & R. Neches. (Eds.), Production system models of learning and development. Cambridge, MA: MIT Press.

Langley, P., Gennari, J.H.& Iba, W. (1987). Hill-climbing theories of learning. Proceedings of the International Workshop on Machine Learning.

Langley, P.& Neches, N. (1981). PRISM users manual (Technical Report). Pittsburgh, PA: Carnegie-Mellon University, Department of Computer Science.

Langley, P., Simon, H.A., Bradshaw, G.L.& Zytkow, J.M. (1987). Scientific discovery: computational explorations of the creative process. Cambridge,MA: MIT Press.

Langley, P., Zytkow, J.M., Simon, H.A., & Bradshaw, G.L., (1986). The search for regularity: four aspects of scientific disovery. In R.S. Michalski, J. Carbonell, & T.M. Mitchell. (Eds.), Machine learning. An arificial intelligence approach(Vol. 2). Los Altos, CA: Kaufmann.

Larkin, J.H. (1983). Mechanisms of effective problem representation in physics. (Resarch Report). Pittsburgh, PA: Carnegie Mellon University.

Larkin, J.H., & Simon, H.A. (1987). Why a diagram is (sometimes) worth ten thousand words. Cognitive Science, 11, 65-99.

Lenat, D.B. (1982). AM: Discovery in mathematics and heuristic search (Dissertation). Stanford, CA: Department of Computer Science, Stanford University.

Levine, M. (1966). Hypothesis behavior by humans during discrimination learning. Journal of Experimental Psychology, 71, 331-338.

Lewis, C. (1978). Production system models of practice effects (Dissertation). Department of Psychology, University of Michigan.

Lindsay, R.K., Buchanan, B.G.& Lederberg, J. (1980). Applications of artificial intelligence for organic chemistry. New York: McGraw Hill.

Medin, D.L., & Smith, E.E. (1984). Concept and concept formation. Annual Review of Psychology, 35, 113-138.

Medin, D.L., Wattenmaker, W.D., & Michalski, R.S. (1987). Constraints and preferences in inductive learning: An experimental study of human and machine performance. Cognitive Science, 11, 299-339.

Mervis, C.B., & Rosch, E. (1981). Categorization of natural objects. Annual Review of Psychology, 32, 89-115.

Michalski, R.S. (1983). A theory and methodology of inductive learning. Artificial Intelligence, 20, 111-161.

Michalski, R.S., Carbonell, J.& Mitchell, T.M.(Eds.) (1983). Machine learning. An artificial intelligence approach. Palo Alto: Tioga Press.

Michalski, R.S., Carbonell, J.& Mitchell, T.M.(Eds.) (1986). Machine learning. An artificial intelligence approach (Vol. 2). Los Altos, CA: Kaufmann.

Mitchell, T.M. (1977). Version Space: A candidate elimination approach to rule learning. Proceedings of the International Joint Conference on Artificial Intelligence.

Mynatt, C.R., Doherty, M.E., & Tweney, R.D. (1977). Confirmation bias in a simulated research environment: An experimental study of scientific inference. Quarterly Journal of Experimental Psychology, 29, 85-95.

Mynatt, C.R., Doherty, M.E., & Tweney, R.D. (1978). Consequences of confirmation and disconfirmation in a simulated research environment. Quarterly Journal of Experimental Psychology, 30, 395-406.

Neches, R., Langley, P., & Klahr, D., (1987). Learning, development, and production systems. In D. Klahr, P. Langley, & D. Neches. (Eds.), Production system models of learning and development. Cambridge,MA: MIT Press.

Neisser, U., & Weene, P. (1962). Hierarchies in concept attainment. Journal of Experimental Psychology, 64, 640-645.

Neves, D.M., & Anderson, J.R., (1981). Knowledge compilation: Mechanisms for the automatization of cognitive skills. In J.R. Anderson. (Ed.), Cognitive skills and their acquisition. Hillsdale, NJ: Erlbaum.

Newell, A. (1967). Studies in Problem Solving: Subject 3 on the crypt-arithmetic task Donald + Gerald = Robert (Technical report). Pittsburgh, PA: Center for the study of human information processing, Carnegie-Mellon University.

Newell, A.& McDermott, J. (1975). PSG manual (Technical Report). Pittsburgh, PA: Department of Computer Science, Carnegie-Mellon University.

Newell, A., & Rosenbloom, P.S., (1981). Mechanisms of skill acquisition and the law of practice. In J.R. Anderson. (Ed.), Cognitive skills and their acquisition. Hillsdale,NJ: Erlbaum.

Newell, A.& Simon, H.A. (1972). Human problem solving. Englewood Cliffs, NJ: Prentice-Hall.

Ohlsson, S.& Langley, P. (1984). PRISM. Tutorial and manual (Technical report). Pittsburgh, PA: Carnegie-Mellon University.

Opwis, K., Stumpf, M.& Spada, H. (1987). PRISM: Einführung in die Theorie und Anwendung von Produktionssystemen. (Technical Report). Freiburg, FRG: Psychological Institute, University of Freiburg.

Peirce, C.S.(1931-1958). Collected papers, 8 vols. Edited by C. Hartshorne, P. Weiss, and A. Burks. Cambridge, MA: Harvard University Press.

Plötzner, R.& Opwis, K. (1987). Modeling discrimination learning in a production system framework. (Tehnical Report). Freiburg, FRG: Psychological Institute, University of Freiburg.

Post, E.L. (1943). Formal reductions of the general combinatorial decision problem. Ajm, 65, 197-268.

Pylyshyn, Z.W., (1984). Computation and cognition. Cambridge, MA: MIT Press.

Reed, S.K., Ernst, G.W., & Banerji, R. (1974). The role of analogy in transfer between similar problem states. Cognitive Psychology, 6, 436-450.

Reimann, P., (1985). Die Funktion der Hypothese im Lernprozeß. In W. Twellmann. (Ed.), Handbuch für Schule und Unterricht, Band 8.1. Düsseldorf, FRG: Schwann Verlag.

Reimann, P. (1988a). Refract - a microworld for refraction phenomena. Program documentation (Technical Report). Freiburg, FRG: Psychological Institute, University of Freiburg.

Reimann, P. (1988b). Individual differences in hypothesis-guided discovery learning. Observations and a protocol analysis. (Technical Report). Freiburg, FRG: Psychological Institute, University Freiburg.

Reimann, P. (1988c). A menu-interface to the PRISM production system shell (unpublished technical paper). Freiburg, FRG: Research Group on Cognitive Systems, Psychological Institute, University of Freiburg.

Reimann, P. (1988d). A frame-based programming tool for rapid prototyping in the PRISM shell (unpublished technical paper). Freiburg: Research Group on Cognitive Systems, Psychological Institute, University of Freiburg.

Reimann, P. (1989). Towards general knowledge diagnosis systems for student and user modelling. In H. Mandl, E. de Corte, N. Bennett, & H.F. Friedrich (Eds.), Learning and instruction. European research in an international context. (in press).

Shrager, J. (1985). Instructionless learning: discovery of the mental model of a complex device. Pittsburgh, PA: Department of Psychology, Carnegie-Mellon University.

Shrager, J. (1987). Instructionless learning about a complex device. Machine Learning, 2, 247-276.

Shultz, R.R., & Kestenbaum, N.R. (1985). Causal reasoning in children. Annals of Child Development, 2, 159-249.

Shute, V., Glaser, R., & Raghavan, K., (1988). Inference and discovery in an exploratory environment. In P.L. Ackerman, R.J. Sternberg, & R. Glaser. (Eds.), Learning and individual differences. (to appear). San Francisco: Freeman.

Siklossy, L. (1968). Natural language learning by computer (Doctoral dissertation). Pittsburgh, PA: Department of Computer Science, Carnegie-Mellon University.

Simon, H.A. (1975). The functional equivalence of problem solving skills (Quoted after the reprint in: Simon, H.A. (1979): Models of thought). New Haven and London: Yale University Press.

Simon, H.A. (1977). Models of discovery and other topics in the methods of science. Dortrecht: Reidel.

Simon, H.A. (1981). The sciences of the artificial. 2nd ed. Cambridge,MA: MIT Press.

Simon, H.A., & Hayes, J.R. (1976). The understanding process: Problem isomorphs. Cognitive Psychology, 8, 165-194.

Simon, H.A., & Kotovsky, K. (1963). Human acquisition of concepts for sequential patterns. Psychological Review, 70, 534-546.

Simon, H.A., & Lea, G., (1974). Problem solving and rule induction: a unified view. In L.W. Gregg. (Ed.), Knowledge and cognition. Hillsdale, NY: Erlbaum.

Smith, R.B. (1986). The alternate reality kit. An animated environment for creative interactive simulations. Dallas, TX: IEEE Computer Society Workshop On Visual Languages.

Spada, H., Reimann, P., & Häusler, B., (1983). Hypothesenerarbeitung und Wissensaufbau beim Schüler. In L. Kötter, & H. Mandl. (Eds.), Kognitive Prozesse und Unterricht. (Jahrbuch für Empirische Erziehungswissenschaft). Düsseldorf: Schwann.

Stumpf, M., Branskat, S., Herderich, C., Newen, A., Opwis, K., Plötzner, R., Schult, R.& Spada, H. (1988). The graphical user interface of DiBi, a microworld for collision phenomena. (Technical Report). Freiburg: Dept. of Psychology, University of Freiburg.

Tweney, R.D, Doherty, M.E.& Mynatt, C.R.(Eds.) (1981). On Scientific Thinking. New York: Columbia University Press.

Utgoff, P.E. (1986). Machine learning of inductive bias. Boston: Kluwer.

VanLehn, K. (1988). Problem solving and cognitive skill acquisition (Technical Report). Pittsburgh, PA: Departments of Psychology and Computer Science, Carnegie-Mellon University.

VanLehn, K., Brown, J.S., & Greeno, J.G., (1984). Competitive argumentation in computational theories of cognition. In W. Kintsch, J. Miller, & P. Polson. (Eds.), Methods and tactics in cognitive science. Hillsdale,NJ: Erlbaum.

VanLehn, K.& Garlick, S. (1987). Cirrus: an automated protocol analysis tool. (Technical Report). Pittsburgh, PA: Depart. of Psychology, Carnegie-Mellon University.

Wason, P.C. (1960). On the failure to eliminate hypotheses in a conceptual task. Quarterly Journal of Experimental Psychology, 12, 129-140.

Wason, P.C. (1968). Reasoning about a rule. Quarterly Journal of Psychology, 20, 273-281.

Wason, P.C.& Johnson-Laird, P.N. (1972). Psychology of reasoning: structure and content. Cambrdige,MA: Harvard University Press.

Waterman, D.A. (1975). Adaptive production systems. Tbilisi, USSR: Proceedings of the Fourth International Joint Conference on Artificial Intelligence.

Wetherick, N.E. (1962). Eliminative and and enumerative behaviour in a conceptual task. Quarterly Journal of Experimental Psychology, 14, 246-249.

Winston, P.H. (1970). Learning structural descriptions from examples (Technical report). Boston, MA: Massachusetts Institute of Technology.

APPENDIX IV.1

Experiment Design Sequences For All Subjects (Ray Experiments Only)

S1

Exp.#	Substance	Radius	ODist	Alpha
3	G	pl	-100	-10
5	D	pl	-100	**-20**[1]
7	D	**-120**	**-50**	**10**
13	D	**pl**	**-100**	10
14	G	pl	**-200**	**15**
15	G	pl	-200	**20**
16	D	pl	**-150**	**15**
17	F	pl	-150	15
18	F	pl	**-100**	**20**
19	G	**120**	**-150**	**15**
20	G	120	**-100**	**10**
21	G	**100**	**-200**	**15**
22	G	**140**	**-50**	**-10**
23	F	**120**	**-150**	**10**
24	F	120	-150	10
25	D	**100**	**-250**	10
26	D	**140**	**-150**	**15**
27	G	**-120**	-150	**10**
28	G	**-100**	**-50**	10

S2

Exp.#	Substance	Radius	ODist	Alpha
1	G	pl	-150	20
2	G	pl	-150	**-20**
3	F	pl	-150	**20**
4	F	pl	-150	**-20**
5	D	pl	-150	**20**
6	G	**100**	-150	20
7	F	100	-150	20
8	D	100	-150	20
9	G	**-100**	-150	20
10	F	-100	-150	20
11	D	-100	-150	20
12	G	-100	**-250**	20
13	G	**pl**	-250	20
14	G	**100**	-250	**15**
15	G	100	**-150**	15

---

[1] Variable values that changed from one experiment to the next are printed in boldface.

S3

1	F	100	-150	-15
2	D	pl	-100	10
6	G	-120	-200	-20
7	D	-100	-50	10
8	G	100	-150	-15
9	F	100	-100	15
10	G	-140	-150	20
11	F	-120	-200	-10
12	F	-100	-50	10
16	G	100	-50	-10
17	D	120	-50	10
18	G	-120	-150	20

S5

Exp.#	Substance	Radius	ODist	Alpha
1	G	pl	-150	15
2	G	pl	-100	-20
3	F	-120	-150	-15
4	F	-140	-50	15
6	G	100	-100	10
7	G	pl	-250	20
8	F	-140	-150	-20
9	F	pl	-200	10
10	G	pl	-150	10
13	G	pl	-100	-10
15	G	pl	-150	10
16	G	pl	-200	15
20	D	pl	-200	10
21	F	pl	-150	20
22	G	-140	-200	-10
23	G	-140	-100	-10

S7

1	D	100	-200	-10
2	D	**-100**	-200	-10
3	D	**140**	-200	**10**
4	D	**100**	-200	**15**
5	D	100	-200	**10**
6	D	**140**	-200	**-10**
7	**F**	100	-200	**10**
8	F	100	-200	**15**
9	**G**	100	-200	**10**
10	G	100	-200	**15**
11	**F**	**140**	-200	**10**
12	F	**120**	-200	10
13	**D**	**100**	-200	10
14	D	**120**	-200	10
15	**F**	**140**	**-50**	10
16	**D**	140	**-200**	10
20	D	**-100**	-200	10
21	**G**	-100	-200	10
24	**D**	**pl**	-200	10
25	D	pl	-200	**-10**
26	D	pl	**-250**	**-15**

S8

1	G	pl	-150	10
2	**F**	pl	-150	**-10**
3	**D**	pl	-150	-10
4	**G**	pl	**-50**	-10
5	G	**120**	**-150**	-10
6	**F**	120	-150	-10
7	**D**	120	-150	-10
8	**G**	120	**-50**	-10
9	G	**100**	**-150**	-10
10	G	**120**	-150	**20**
11	G	**-120**	-150	**-10**
12	**F**	-120	-150	-10
13	**D**	**120**	-150	-10
14	D	**-120**	**-200**	-10
15	**G**	**-100**	**-150**	-10
16	G	**-120**	-150	**20**

S10

1	F	pl	-200	10
2	F	pl	-200	10
3	F	pl	-200	10
4	**G**	pl	-200	10
5	G	pl	-200	10
6	G	pl	-200	10
7	G	pl	-200	**-10**
8	G	pl	-200	-10
9	G	pl	**-150**	**15**
10	G	pl	**-200**	15
11	G	pl	-200	**20**
12	G	pl	-200	**10**
13	G	pl	-200	**15**
14	**F**	pl	-200	15
15	F	pl	-200	15
16	F	pl	**-50**	**-15**
24	**G**	**100**	**-100**	10
25	G	100	-100	10
26	**F**	100	-100	10
27	**D**	**pl**	-100	10
28	D	pl	-100	**-10**
29	D	**-100**	-100	**10**
30	**G**	**-120**	**-250**	**15**
31	G	**-100**	**-200**	**20**

S11

1	G	pl	-200	15
2	**F**	pl	-200	15
3	**D**	pl	-200	15
4	**G**	**100**	-200	15
5	**F**	100	-200	15
6	**D**	100	-200	15
7	**G**	**-100**	-200	15
8	**F**	-100	-200	15
9	**D**	-100	-200	15
27	**G**	**100**	**-50**	15

APPENDIX VI.1

PRISM Code for Model HDD-SH

This appendix contains the complete production system code for HDD-SH. Supporting Lisp code is not included. The architectural specifications and the production rules of the program are bound to InterLisp variables as a means to make them accessable for editing and to save them as program files. On a level closer to their semantics, they are bundled into frames. Note that the version of HDD-SH printed here does not correspond completely to the description given in Chapter 6. There, I introduced the memory PHEN for purposes of a clearer presentation only in the context of the program HDD-SHOP. Actually, HDD-SH as described here does have the same architecture as HDD-SHOP, i.e., does have the memory PHEN and its consideration rules already, but the Bucket Brigading mechanism is turned off.

All variables that hold PRISM code specifying archtitecture end with the extension '.ARC'. Variables that refer to PRISM structures specifying the declarative content of memories end with extension '.CNT'. All the other variables point to production rule groups.

A PRISM rule has the following general syntax:

(<production name>
     ((<condition>)
      (<condition>)
       ...
)
     -->
     ((<action>)
      (<action>)
       ...
)
).

The following elements have special meaning:

- \#   followed by a number refers to the position of a condition element. The respective condition element is substituted by the interpreter.

- $ Precedes a call to a PRISM memory function.

- *   Precedes a call to a LISP predicate in the condition side.

- &   Precedes a call to a LISP function in the action side.

```
{FSO:COGNITIVE SYSTEMS:PSYCHOLINSTITUT}<REIMANN>HDD>HDD-SH.;1     4-Mar-89 12:53:44


(FILECREATED " 4-Mar-89 12:53:40" {FSO:COGNITIVE% SYSTEMS:PSYCHOLINSTITUT}<REIMANN>HDD>
HDD-SH.;1

        changes to:  (VARS HDD-SHCOMS))


(PRETTYCOMPRINT HDD-SHCOMS)

(RPAQQ HDD-SHCOMS ((ADDVARS (*MODULES* ENV-0.ARC ENVSpread.Arc WM-0.ARC CONTROL-0.ARC PHEN-1.ARC
                            HM2.ARC DISCRIM-0.ARC GEN-0.ARC HouseHold.PM GenDesign.2
                            Predict&Compare.2 ModifyH.2 GenH.2 TrendDetector.1
                            FunctionInducer.1 WM-1.CNT HM1.CNT))
        (PROP FRAME LEARNER-3)
        (COMS (* Architectural Definitions)
              (VARS ENV-0.ARC ENVSpread.Arc WM-0.ARC CONTROL-0.ARC PHEN-1.ARC HM2.ARC DISCRIM-0.ARC))
        (COMS (* Productions)
              (VARS HouseHold.PM GenDesign.2 Predict&Compare.2 ModifyH.2 GenH.2 TrendDetector.1
                    FunctionInducer.1))
        (COMS (* Initial Contents of memories WM and HM)
              (VARS WM-1.CNT HM1.CNT))))

(ADDTOVAR *MODULES* ENV-0.ARC ENVSpread.Arc WM-0.ARC CONTROL-0.ARC PHEN-1.ARC HM2.ARC DISCRIM-0.ARC
                    GEN-0.ARC HouseHold.PM GenDesign.2 Predict&Compare.2 ModifyH.2 GenH.2

                    TrendDetector.1 FunctionInducer.1 WM-1.CNT HM1.CNT)

(PUTPROPS LEARNER-3 FRAME [LEARNER-3 (DOC (* * A learner who prefers more simple phenomenon
                                              predictions over more complex ones))
                          (A-KIND-OF (VALUE NO.BUCK.BRIGD.))
                          [ENV-Arc (VALUE (ENV ARC (ENV-0.ARC ENVSpread.Arc]
                          (WM-Arc (VALUE (WM ARC WM-0.ARC)))
                          (CONTROL-Arc (VALUE (CONTROL ARC CONTROL-0.ARC)))
                          (PHEN-Arc (VALUE (PHEN ARC PHEN-1.ARC)))
                          [HM-Arc (VALUE (HM ARC HM2.ARC)
                                         (DOC (* changed to L2]
                          (DISCRIMINATION (VALUE (DISCRIMINATE ARC DISCRIM-0.ARC)))
                          (GENERALIZATION (VALUE (GENERALIZE ARC GEN-0.ARC)))
                          (CONTROL-Content (VALUE (CONTROL PROC HouseHold.PM)))
                          (GenDesign-Content (VALUE (CONTROL PROC GenDesign.2))
                                             (DOC (* Changed L2)))
                          (Predict-Content (VALUE (CONTROL PROC Predict&Compare.2)))
                          (ModifyH-Content (VALUE (CONTROL PROC ModifyH.2)))
                          [GenH-Content (VALUE (PHEN PROC (GenH.2 TrendDetector.1
                                                                  FunctionInducer.1]
                          (ENV-Content (VALUE NIL))
                          (WM-Content (VALUE (WM DECL WM-1.CNT)))
                          (HM-Content (VALUE (HM PROC HM1.CNT)])


(* Architectural Definitions)


(RPAQQ ENV-0.ARC (PROGN (* * set up an external memory ENV)
                        (CREATE-COMPONENT ENV INSTANCE-OF DECLARATIVE-MEMORY ATTRIBUTES (ACTIVATION)

                                          DEFAULT-VALUES
                                          (1.0)
                                          REVISED-VALUES
                                          (NEW)
                                          SYNTAX
                                          (Exp))
                        (printout T T "Created Architecture for ENV" T)))

(RPAQQ ENVSpread.Arc [PROGN (* * Define a retrieval process for ENV)
                            (CREATE-COMPONENT SPREAD-TO-LIMIT INSTANCE-OF PROPAGATION-PROCESS
                                              PROCESSES ((MODIFY-AND-ADD-TO WM ELEMENT))
                                              DEFAULT-AMOUNT 1.0 SPREAD-FROM-ELEMENT
                                              ((NEW-AMOUNT (FTIMES AMOUNT .5))
                                               (NEW-AMOUNT (FQUOTIENT AMOUNT (NUMBER-OF-SYMBOLS)))
                                               (BLOCK-ELEMENT)
                                               (SPREAD-TO-SYMBOLS)
                                               (UNBLOCK-ELEMENT))
                                              SPREAD-TO-SYMBOLS
                                              ((NEW-AMOUNT (FTIMES AMOUNT (NUMBER-OF-TIMES)))
                                               (SPREAD-FROM-SYMBOL))
                                              SPREAD-FROM-SYMBOL
```

```
                              ((NEW-AMOUNT (FQUOTIENT AMOUNT (SUM-OF ENV-ACTIVATION)
                                                                                    ))
                               (BLOCK-SYMBOL)
                               (SPREAD-TO-ELEMENTS)
                               (UNBLOCK-SYMBOL))
                              SPREAD-TO-ELEMENTS
                              ((NEW-AMOUNT (FTIMES AMOUNT (ENV-ACTIVATION ELEMENT)))

                               (IS-AMOUNT? (GREATERP AMOUNT .05))
                               (NEW-AMOUNT (MIN AMOUNT 5.0))
                               (STORE-AMOUNT AMOUNT)
                               (SPREAD-FROM-ELEMENT])

(RPAQQ WM-0.ARC (PROGN (* * Create a recency based working memory)
                       (MD WM ATTRIBUTES (RECENCY)
                              DEFAULT-VALUES
                              ((WM-COUNT+))
                              REVISED-VALUES
                              (NEW))))

(RPAQQ CONTROL-0.ARC [PROGN (* * Set up memory CONTROL %. Prefer more recent entries and more
                                 specific entries)
                             (CREATE-COMPONENT CONTROL INSTANCE-OF PROCEDURAL-MEMORY MATCHES-AGAINST
                              WM EVERY-CYCLE ((ORDER-BY-FIRST-ELEMENT WM-RECENCY)
                               (SELECT-BEST GREATERP)
                               (ORDER-BY-SUMMED-ELEMENTS WM-RECENCY)
                               (SELECT-ONE-BEST GREATERP)
                               (FIRE-PRODUCTIONS)
                               (REFRACT-FIRED))
                              EVERY-FIRING
                              ((PRINT-FIRING)
                               (PRINT-NAME)
                               (MY-PRODUCTION-WATCHER)
                               (EVALUATE])

(RPAQQ PHEN-1.ARC [PROGN (* * Architecture of PHEN, the memory that holds rules for phenomenon
                              induction. - Uses strength to select consideration rules)
                          (CREATE-COMPONENT PHEN INSTANCE-OF PROCEDURAL-MEMORY MATCHES-AGAINST WM
                                 ATTRIBUTES (STRENGTH)
                                 DEFAULT-VALUES
                                 (1.0)
                                 REVISED-VALUES
                                 (OLD)
                                 EVERY-CYCLE
                                 ((ORDER-BY-PRODUCTION PHEN-STRENGTH)
                                  (SELECT-BEST GREATERP)
                                  (FIRE-PRODUCTIONS)
                                  (REFRACT-FIRED))
                                 EVERY-FIRING
                                 ((PRINT-FIRING)
                                  (PRINT-NAME)
                                  (MY-PRODUCTION-WATCHER)
                                  (EVALUATE])

(RPAQQ HM2.ARC (PROGN (* * Hypothesis Memory - Productions are selected according to strength)
                       (CREATE-COMPONENT HM INSTANCE-OF PROCEDURAL-MEMORY MATCHES-AGAINST WM
                              ATTRIBUTES (STRENGTH)
                              DEFAULT-VALUES
                              (1.0)
                              REVISED-VALUES
                              ((FPLUS OLD 1.0))
                              EVERY-CYCLE
                              ((ORDER-BY-PRODUCTION HM-STRENGTH)
                               (SELECT BEST GREATERP)
                               (ORDER-BY-PRODUCTION HM-STRENGTH)
                               (SELECT-RANDOMLY)
                               (FIRE-PRODUCTIONS)
                               (REFRACT-FIRED)
                               (DELETE-FROM WM (GOAL: predict))
                               (CALL CONTROL))
                              EVERY-FIRING
                              ((DIVIDER)
                               (PRINT-FIRING)
                               (PRINT-NAME)
                               (EVALUATE))
                              DEFAULT-ACTIONS NIL SYNTAX NIL GRAPHFLG T)
```

```
                             (* * Definition of the "Forgetting" Process)
                             (CREATE-COMPONENT HM-VANISH INSTANCE-OF TRANSFER-PROCESS TEST
                                               (LESSP (STRENGTH ELEMENT)
                                                      .5)
                                               PROCESSES
                                               ((PRINT-TRANSFER)
                                                (WRITECR UNBUILDING ELEMENT)
                                                (UNBUILD-FROM HM ELEMENT)
                                                (HDD.DribbleTransfer HM-VANISH ELEMENT)))
                             (* The current lower limit corresponds to two decreases starting from the
                                initial value 1.0)
                             (printout T T "Created Transfer Process HM-VANISH " T)))

(RPAQQ DISCRIM-0.ARC (PROGN (* * Definition of the discrimination components DISCRIMINATE,
                                 ASSIGN-CREDIT, INCREASE-ATTRIBUTE-VALUE and DECREASE-ATTRIBUTE-VALUE)

                            (CREATE-COMPONENT DISCRIMINATE INSTANCE-OF DISCRIMINATION-PROCESS
                                              WORKING-MEMORY WM LEARNING-ACTIONS ((
                                                POSITIVE-CONDITIONS))
                                              SYMBOL-REPLACEMENT OFF LIMIT 1 SIGNIFICANT-SYMBOLS
                                              (Exp GOAL: Substance Radius Alpha Theta1 Gamma Theta2

                                                ODist predict)
                                              SIGNIFICANT-SYMBOLS-TESTS
                                              (EXP←LABEL←P)
                                              PRESELECTION T SELECTION-SYMBOLS (Exp))
                            (CREATE-COMPONENT ASSIGN-CREDIT INSTANCE-OF DESIGNATION-PROCESS MEMORY
                                              WM PROCESS NIL)
                            (CREATE-COMPONENT INCREASE-ATTRIBUTE-VALUE INSTANCE-OF REVISION-PROCESS

                                              PROCESS INCREASE ATTRIBUTE STRENGTH FUNCTOR PLUS
                                              FACTOR 1.0)
                            (CREATE-COMPONENT DECREASE-ATTRIBUTE-VALUE INSTANCE-OF REVISION-PROCESS

                                              PROCESS DECREASE ATTRIBUTE STRENGTH FUNCTOR TIMES
                                              FACTOR .25)))
```

(* Productions)

```
(RPAQQ HouseHold.PM [BUILD-IN CONTROL (* * HOUSEHOLD RULES * *)
                                (* * The focus rules manage the content of WM according to the
                                   content of (ExpsInFocus: <> <>))
                                [HH.Focus ((GOAL: focus on experiment =newfocus)
                                           (ExpsInFocus: =oldfocus)
                                           (ExpsInNoteBook: !nb)
                                           (*NOT (*EQUAL =newfocus =oldfocus)))
                                          -->
                                          (($DELETE-FROM WM #2)
                                           ($BIND =moveout (&LDIFFERENCE =oldfocus =newfocus))
                                           ($BIND =movein (&LDIFFERENCE =newfocus =oldfocus))
                                           ($WRITECR "** Removing " =moveout from Focus)
                                           ($HDD.RemoveExpFromWM* =moveout)
                                           ($WRITECR "** Moving " =movein into Focus)
                                           ($HDD.AddExpToWM* =movein =nb)
                                           ($ADD-TO WM (ExpsInFocus: =newfocus)]
                                (HH.FocusDone ((GOAL: focus on experiment !exps))
                                              -->
                                              (($DELETE-GOAL WM #1])
(RPAQQ GenDesign.2 [BUILD-IN CONTROL (* * Rules for designing an experiment * *)
                                [GenDesign2.SetSubGoals ((GOAL: run experiment))
                                                        -->
                                                        (($CALL CONTROL)
                                                         ($DELETE-GOAL WM #1)
                                                         ($ADD-GOAL WM (GOAL: design experiment]
                                (GenDesign0.Stop ((GOAL: design experiment)
                                                  (CurrentExperiment =e1)
                                                  (Last experiment =e1))
                                                 -->
                                                 (($DELETE-GOAL WM #1)
                                                  ($WRITECR No more experiments ... I stop)
                                                  ($STOP)))
                                [GenDesign0.DesignExp ((GOAL: design experiment)
                                                       (CurrentExperiment =e))
                                                      -->
```

```
                                ((SDELETE-GOAL WM #1)
                                 (SDELETE-FROM WM #2)
                                 ($BIND =next (&HDD.NextExperiment =e))
                                 ($ADD-TO WM (CurrentExperiment =next))
                                 ($ADD-GOAL WM (GOAL: get design values)
                                                (GOAL: focus on experiment (=next]
               (GenDesignO.GetValues ((GOAL: get design values)
                                      (CurrentExperiment =e))
                                -->
                                (($HDD.ReadDesignValues =e)
                                 ($DELETE-GOAL WM #1)
                                 ($ADD-GOAL WM (GOAL: predict]))

(RPAQQ Predict&Compare.2 [BUILD-IN CONTROL (* * Rules for deriving a prediction and comparing it
                        with feedback * *)
           [Pred&CompO.MakePrediction ((GOAL: predict))
                                -->
                                (($CALL HM)
                                 ($ADD-GOAL WM (GOAL: evaluate
                                                         prediction]
           [Pred&CompO.EvalPred ((GOAL: evaluate prediction))
                                -->
                                (($DELETE-GOAL WM #1)
                                 ($ADD-GOAL WM (GOAL: compare prediction
                                                         feedback)
                                                (GOAL: get feedback]
           (Pred&CompO.GetFeedbackValues ((GOAL: get feedback)
                                          (ExpsInNoteBook: !first =some)
                                          (CurrentExperiment =e)
                                          (*NOT (*EQUAL =some =e)))
                                -->
                                (($DELETE-FROM WM #2)
                                 ($ADD-TO WM
                                           (ExpsInNoteBook: !first
                                                             =some =e))
                                 ($HDD.ReadFeedbackValues =e)
                                 ($DELETE-GOAL WM #1)))
           (Pred&CompO.GetFeedbackValues1 ((GOAL: get feedback)
                                           (CurrentExperiment =e)
                                           (ExpsInNoteBook:))
                                -->
                                (($DELETE-FROM WM #3)
                                 ($ADD-TO WM (ExpsInNoteBook:
                                                =e))
                                 ($HDD.ReadFeedbackValues =e)
                                 ($DELETE-GOAL WM #1)))
           [Pred&CompO.Compare1 ((GOAL: compare prediction feedback)
                                 (=p ISA: Prediction !form)
                                 (CurrentExperiment =e))
                                -->
                                (($ADD-GOAL WM (GOAL: compare #2 =e]
           [Pred&Comp1.CorrectPred ((GOAL: compare
                                            (=p ISA: Prediction DV: =d IV:
                                                =i Form: =form Value: =v1)
                                            =e)
                                    (Exp =e =d =v2)
                                    (*HDD.NumberAlmostEqualP =v1 =v2)
                                    (FailureCount =n))
                                -->
                                (($WRITECR ***Prediction is correct)
                                 ($DELETE-GOAL WM #1)
                                 ($DELETE-FROM WM (FailureCount =n))
                                 ($ADD-TO WM
                                           (Correct (=p ISA: Prediction
                                                         DV: =d IV: =i
                                                         Form: =form
                                                         Value: =v1))
                                           (FailureCount 0]
           (* Changed: Ihis rule keeps a failure count)
           (Pred&Comp1.WrongPred ((GOAL: compare
                                          (=p ISA: Prediction DV: =d IV: =i
                                              Form: =form Value: =v1)
                                          =e)
                                  (Exp =e =d =v2)
                                  (*NOT (*HDD.NumberAlmostEqualP =v1 =v2))
                                  (FailureCount =n))
                                -->
```

```
                                    (($DELETE-GOAL WM #1)
                                     ($DELETE-FROM WM (FailureCount =n))
                                     ($ADD-TO WM
                                              (DifferencePredFeedb: DV: =d
                                                                   Predicted: =v1

                                                                   FbValue: =v2)

                                              (FailureCount (&ADD1 =n)))
                                     ($WRITECR ***Prediction is wrong)))
                       [Pred&Comp0.CompareDone ((GOAL: compare prediction feedback)
                                                (Hypothesis: !x))
                                               -->
                                               (($HDD.RemoveFirstsFromWM Hypothesis:)
                                                ($DELETE-GOAL WM #1)
                                                ($ADD-GOAL WM (GOAL: eval hypotheses)]
                       (Pred&Comp0.NoH ((GOAL: compare prediction feedback)
                                        (<NOT> (Hypothesis: !x)))
                                       -->
                                       (($DELETE-GOAL WM #1)
                                        ($ADD-GOAL WM (GOAL: generate hypotheses)])

(RPAQQ ModifyH.2 [BUILD-IN CONTROL (* * Rules that trigger discrimination and other modifications
                        of hypotheses * *)
                       [ModifyH2.CorrectH ((GOAL: eval hypotheses)
                                           (Correct (=p ISA: Prediction DV: =d IV: =i Form:
                                                        =form Value: =v1)))
                                          -->
                                          (($WRITECR ***Hypothesis is correct, I strengthen the

                                                        rule)
                                           ($HDD.DribblePrediction (=p ISA: Prediction DV: =d
                                                                       IV: =i Form: =form
                                                                       Value: =v1)
                                                                   (&GET-RULE-NAME =p)
                                                                   correct)
                                           ($STORE-CONDITIONS ($GET-CREDIT (=p ISA: Prediction
                                                                               DV: =d IV: =i
                                                                               Form: =form
                                                                               Value: =v1)))
                                           ($INCREASE-ATTRIBUTE-VALUE ($GET-RULE-NAME =p))
                                           ($DELETE-FROM WM #2)
                                           ($DELETE-FROM WM
                                                          (=p ISA: Prediction DV: =d IV: =i
                                                              Form: =form Value: =v1]
                       (* BB turned off)
                       [ModifyH2.WrongPred ((GOAL: eval hypotheses)
                                            (=p ISA: Prediction DV: =d IV: =i Form: =form
                                                Value: =v1)
                                            (DifferencePredFeedb: DV: =d Predicted: =v1
                                                                  FbValue: =v2))
                                           -->
                                           (($WRITECR ***Prediction is wrong...I try to find
                                                         new conditions)
                                            ($HDD.DribblePrediction (=p ISA: Prediction DV: =d
                                                                        IV: =i Form: =form
                                                                        Value: =v1)
                                                                    (&GET-RULE-NAME =p)
                                                                    wrong)
                                            ($DECREASE-ATTRIBUTE-VALUE (&GET-RULE-NAME =p))
                                            ($DISCRIMINATE (=p ISA: Prediction DV: =d IV: =i
                                                               Form: =form Value: =v1))
                                            ($HM-VANISH HM)
                                            ($DELETE-FROM WM #2 #3)
                                            ($ADD-TO WM (A hypothesis was wrong]
                       (* BB turned off)
                       [ModifyH0.CallGenerator ((GOAL: eval hypotheses)
                                                (A hypothesis was wrong)
                                                (<NOT> (Correct =x))
                                                (<NOT> (DifferencePredFeedb: !rest)))
                                               -->
                                               (($DELETE-GOAL WM #1 #2)
                                                ($ADD-GOAL WM (GOAL: generate hypotheses]
                       (ModifyH0.NextExp ((GOAL: eval hypotheses)
                                          (<NOT> (Correct =x))
                                          (<NOT> (A hypothesis was wrong)))
                                         -->
                                         (($DELETE-GOAL #1)
```

```
                                  ($ADD-GOAL WM (GOAL: run experiment])
(RPAQQ GenH.2 [BUILD-IN CONTROL (* * Rules That Coordinate the Creation of New Hypotheses * *)
              [GenH2.TurnOnConsiderRules ((GOAL: generate hypotheses)
                                         (CurrentExperiment =e)
                                         (FailureCount =n)
                                         (*GEQ =n 4))
                                         -->
                                         (($DELETE-FROM WM (FailureCount =n))
                                          ($ADD-TO WM (FailureCount 0))
                                          ($MASK (&HDD.ResetAttribute STRENGTH PHEN 1.0]
              [GenH1.ShiftFocus ((GOAL: generate hypotheses)
                                (CurrentExperiment =e)
                                (ExpsInNoteBook: !first =last =e))
                                -->
                                (($DELETE-FROM WM #1)
                                 ($ADD-GOAL WM (GOAL: consider variables)
                                              (GOAL: focus on experiment (=last =e]
              (GenH1.CallTrendDetector1 ((GOAL: consider variables))
                                        -->
                                        (($CALL PHEN)))
              (GenH1.CallTrendDetector2 ((GOAL: consider variables)
                                        (No more variable considerations))
                                        -->
                                        (($DELETE-GOAL WM #1)
                                         ($DELETE-FROM WM #2)
                                         ($ADD-GOAL WM (GOAL: find trends))
                                         ($CALL PHEN)))
              [GenH1.TrendDetectorDone ((GOAL: find trends)
                                       (No more trends)
                                       (Relation: !rest)
                                       (CurrentExperiment =e))
                                       -->
                                       (($DELETE-GOAL WM #1)
                                        ($DELETE-FROM WM #2)
                                        ($ADD-GOAL WM (GOAL: consider functions)
                                                     (GOAL: focus on experiment (=e]
              [GenH1.Done2 ((GOAL: find trends)
                           (No more trends)
                           (<NOT> (Relation: !rest)))
                           -->
                           (($DELETE-GOAL WM #1)
                            ($DELETE-FROM WM #2)
                            ($HDD.RemoveFirstsFromWM Consider:)
                            ($HDD.RemoveFirstsFromWM Relation:)
                            ($ADD-GOAL WM (GOAL: run experiment]
              (GenH1.CallFunctionInducer1 ((GOAL: consider functions))
                                          -->
                                          (($CALL PHEN)))
              (GenH1.CallFunctionInducers ((GOAL: consider functions)
                                          (No more function considerations))
                                          -->
                                          (($DELETE-GOAL WM #1)
                                           ($DELETE-FROM WM #2)
                                           ($ADD-GOAL WM (GOAL: find functions))
                                           ($CALL PHEN)))
              [GenH1.Done1 ((GOAL: find functions)
                           (No more functions)
                           (CurrentExperiment =e))
                           -->
                           (($DELETE-GOAL WM #1)
                            ($DELETE-FROM WM #2)
                            ($HDD.RemoveFirstsFromWM Function:)
                            ($HDD.RemoveFirstsFromWM Relation:)
                            ($HDD.RemoveFirstsFromWM Consider:)
                            ($ADD-GOAL WM (GOAL: run experiment]
              (GenH1.BlockGeneration ((GOAL: generate hypotheses)
                                     (CurrentExperiment exp-1))
                                     -->
                                     (($DELETE-GOAL WM #1)
                                      ($WRITECR "*** I removed " #1
                                                " since this is the first experiment")
                                      ($ADD-GOAL WM (GOAL: run experiment])
(RPAQQ TrendDetector.1 [BUILD-IN PHEN (* * Variable consideration rules * *)
                       [TREND.Substance ((GOAL: consider variables)
                                        (<NOT> (Consider: Variable Substance)))
                                        -->
```

```
                            (($ADD-TO WM (Consider: Variable Substance]
[TREND.ConsiderAlpha ((GOAL: consider variables)
                      (<NOT> (Consider: Variable Alpha)))
                     -->
                     (($ADD-TO WM (Consider: Variable Alpha]
[TREND.ConsiderODist ((GOAL: consider variables)
                      (<NOT> (Consider: Variable ODist)))
                     -->
                     (($ADD-TO WM (Consider: Variable ODist]
[TREND.ConsiderIDist ((GOAL: consider variables)
                      (<NOT> (Consider: Variable IDist)))
                     -->
                     (($ADD-TO WM (Consider: Variable IDist]
[TREND.ConsiderTheta1 ((GOAL: consider variables)
                       (<NOT> (Consider: Variable Theta1)))
                      -->
                      (($ADD-TO WM (Consider: Variable Theta1]
[TREND.ConsiderTheta2 ((GOAL: consider variables)
                       (<NOT> (Consider: Variable Theta2)))
                      -->
                      (($ADD-TO WM (Consider: Variable Theta2]
[TREND.ConsiderGamma ((GOAL: consider variables)
                      (<NOT> (Consider: Variable Gamma)))
                     -->
                     (($ADD-TO WM (Consider: Variable Gamma]
(TREND.ConsiderDone ((GOAL: consider variables))
                    -->
                    (($ADD-TO WM (No more variable considerations))

                    ($CALL CONTROL)))
(* * A set of Trend detectors * *)
[TREND.NominalChange ((GOAL: find trends)
                      (Dependent =d)
                      (Independent =i)
                      (Nominal =i)
                      (Consider: Variable =i)
                      (Consider: Variable =d)
                      (<NOT> (Relation: TREND-NOMINAL
                                        (=d =i)))
                      (Exp =e1 =i =val4)
                      (Exp =e2 =i =val5)
                      (*NEQ =e1 =e2)
                      (*NEQ =val4 =val5)
                      (Exp =e1 =d =val1)
                      (Exp =e2 =d =val2)
                      (*NOT (*EQP =val1 =val2)))
                     -->
                     (($ADD-TO WM (Relation: TREND-NOMINAL
                                             (=d =i]
[TREND.Direct1 ((GOAL: find trends)
                (Dependent =d)
                (Independent =i)
                (Numeric =i)
                (Consider: Variable =i)
                (Consider: Variable =d)
                (SameType (=i =d))
                (<NOT> (Relation: PROP+ (=d =i)))
                (Exp =e1 =d =val1)
                (Exp =e2 =d =val2)
                (*MYGREATERP =val2 =val1)
                (Exp =e1 =i =val4)
                (Exp =e2 =i =val5)
                (*MYGREATERP =val5 =val4))
               -->
               (($ADD-TO WM (Relation: PROP+ (=d =i))
                            ($WRITECR *** As =i increases, =d
                                          increases]
[TREND.Direct2 ((GOAL: find trends)
                (Dependent =d)
                (Independent =i)
                (Numeric =i)
                (Consider: Variable =i)
                (Consider: Variable =d)
                (SameType (=i =d))
                (<NOT> (Relation: PROP+ (=d =i)))
                (Exp =e1 =d =val1)
                (Exp =e2 =d =val2)
                (*MYLESSP =val2 =val1)
```

173

```
                                    (Exp =e1 =i =val4)
                                    (Exp =e2 =i =val5)
                                    (*MYLESSP =val5 =val4))
                                    -->
                                    (($ADD-TO WM (Relation: PROP+ (=d =i))
                                            ($WRITECR *** As =i decreases =d
                                                        decreases]
                    [TREND.Inverse1 ((GOAL: find trends)
                                    (Dependent =d)
                                    (Independent =i)
                                    (Numeric =i)
                                    (Consider: Variable =i)
                                    (Consider: Variable =d)
                                    (SameType (=i =d))
                                    (<NOT> (Relation: PROP- (=d =i)))
                                    (Exp =e1 =d =val1)
                                    (Exp =e2 =d =val2)
                                    (*MYGREATERP =val2 =val1)
                                    (Exp =e1 =i =val4)
                                    (Exp =e2 =i =val5)
                                    (*MYLESSP =val5 =val4))
                                    -->
                                    (($ADD-TO WM (Relation: PROP- (=d =i))
                                            ($WRITECR *** As =i increases =d
                                                        decreases]
                    [TREND.Inverse2 ((GOAL: find trends)
                                    (Dependent =d)
                                    (Independent =i)
                                    (Numeric =i)
                                    (Consider: Variable =i)
                                    (Consider: Variable =d)
                                    (SameType (=i =d))
                                    (<NOT> (Relation: PROP- (=d =i)))
                                    (Exp =e1 =d =val1)
                                    (Exp =e2 =d =val2)
                                    (*MYLESSP =val2 =val1)
                                    (Exp =e1 =i =val4)
                                    (Exp =e2 =i =val5)
                                    (*MYGREATERP =val5 =val4))
                                    -->
                                    (($ADD-TO WM (Relation: PROP- (=d =i))
                                            ($WRITECR *** As =i decreases =d
                                                        increases]
            (TREND.Done ((GOAL: find trends))
                        -->
                        (($ADD-TO WM (No more trends))
                        ($CALL CONTROL])

(RPAQQ FunctionInducer.1 [BUILD-IN PHEN (* * Set of Function detectors * *)
                    [CONSIDER.Sum ((GOAL: consider functions)
                                    (CurrentExperiment =e)
                                    (Relation: PROP- (=d =i))
                                    [<NOT> (Function: DV: =d IV: =i Form:
                                                    (LAMBDA (IV)
                                                            (ROUNDED.PLUS IV =C]
                                    (Numeric =i)
                                    (Numeric =d))
                                    -->
                                    (($ADD-TO WM
                                            (Consider: Function sum exp =e dv =d iv
                                                    =i]
                    [CONSIDER.Difference ((GOAL: consider functions)
                                    (CurrentExperiment =e)
                                    (Relation: PROP+ (=d =i))
                                    [<NOT> (Function: DV: =d IV: =i Form:
                                                    (LAMBDA (IV)
                                                            (
ROUNDED.DIFFERENCE IV =C]
                                    (Numeric =i)
                                    (Numeric =d))
                                    -->
                                    (($ADD-TO WM
                                            (Consider: Function difference
                                                    exp =e dv =d iv =i]
                    [CONSIDER.Product ((GOAL: consider functions)
                                    (CurrentExperiment =e)
                                    (Relation: PROP- (=d =i))
                                    [<NOT> (Function: DV: =d IV: =i Form:
```

```
{FS0:COGNITIVE SYSTEMS:PSYCHOLINSTITUT}<REIMANN>HDD>HDD-SH.;1    4-Mar-89 12:53:44

                                                    (LAMBDA (iv)
                                                        (ROUNDED.TIMES
                                                            iv =c]
                                (Numeric =i)
                                (Numeric =d))
                                -->
                                (($ADD-TO WM
                                            (Consider: Function product exp =e
                                                    dv =d iv =i]
            [CONSIDER.Quotient ((GOAL: consider functions)
                                (CurrentExperiment =e)
                                (Relation: PROP+ (=d =i))
                                [<NOT> (Function: DV: =d IV: =i Form:
                                                    (LAMBDA (iv)
                                                        (ROUNDED.QUOTIENT
                                                            iv =c]
                                (Numeric =i)
                                (Numeric =d)
                                (Exp =e =d =dval)
                                (*NOT (*EQP =dval 0)))
                                -->
                                (($ADD-TO WM
                                            (Consider: Function quotient exp =e
                                                    dv =d iv =i]
            (ConsiderFunctionsDone ((GOAL: consider functions))
                                -->
                                (($ADD-TO WM (No more function
                                                considerations))
                                ($CALL CONTROL)))
            (* * The function inducers * *)
            (FUNCTION.Sum ((Consider: Function sum exp =e dv =d iv =i)
                            (Exp =e =i =ival)
                            (Exp =e =d =dval))
                            -->
                            (($DELETE-FROM WM #1)
                            [$BIND =F (Function: DV: =d IV: =i Form:
                                                    (LAMBDA (IV)
                                                        (ROUNDED.PLUS
                                                            IV
                                                            (
&ROUNDED.DIFFERENCE =dval =ival]
                            ($ADD-TO WM =F)
                            ($WRITECR "*** I found: " =F)
                            ($HDD.PredictionRule =F HM)))
            (FUNCTION.Difference ((Consider: Function difference exp =e dv
                                                    =d iv =i))
                            (Exp =e =i =ival)
                            (Exp =e =d =dval))
                            -->
                            (($DELETE-FROM WM #1)
                            [$BIND =F
                                (Function: DV: =d IV: =i Form:
                                                    (LAMBDA
                                                        (IV)
                                                        (ROUNDED.DIFFERENCE
                                                            IV
                                                            (&ROUNDED.DIFFERENCE
                                                                =ival =dval]
                            ($ADD-TO WM =F)
                            ($WRITECR "*** I found: " =F)
                            ($HDD.PredictionRule =F HM)))
            (FUNCTION.Product ((Consider: Function product exp =e dv =d iv
                                                    =i)
                            (Exp =e =i =ival)
                            (Exp =e =d =dval))
                            -->
                            (($DELETE-FROM WM #1)
                            [$BIND =F (Function: DV: =d IV: =i Form:
                                                    (LAMBDA
                                                        (iv)
                                                        (ROUNDED.TIMES
                                                            iv
                                                            (&ROUNDED.QUOTIENT
                                                                =dval =ival]
                            ($ADD-TO WM =F)
                            ($WRITECR "*** I found: " =F)
                            ($HDD.PredictionRule =F HM)))
```

```
(FUNCTION.Quotient ((Consider: Function quotient exp =e dv =d iv
                                                                =i)
                     (Exp =e =i =ival)
                     (Exp =e =d =dval)
                     (*NOT (*ZEROP =dval)))
                     -->
                     (($DELETE-FROM WM #1)
                      [$BIND =F (Function: DV: =d IV: =i Form:
                                             (LAMBDA
                                               (iv)
                                               (ROUNDED.QUOTIENT
                                                 iv
                                                 (&ROUNDED.QUOTIENT
                                                   =ival =dval]
                      ($ADD-TO WM =F)
                      ($WRITECR "*** I found: " =F)
                      ($HDD.PredictionRule =F HM)))
(FUNCTION.Done ((GOAL: find functions))
                -->
                (($ADD-TO WM (No more functions))
                 ($CALL CONTROL])
```

(* Initial Contents of memories WM and HM)

```
(RPAQQ WM-1.CNT (PROGN (* * Standard information assumed to be available right from the start * *)
                       (ADD-TO WM (Independent Substance)
                                  (Independent Radius)
                                  (Independent ODist)
                                  (Independent Alpha)
                                  (Independent Theta1)
                                  (Dependent IDist)
                                  (Dependent Theta2)
                                  (Dependent Gamma)
                                  (Nominal Substance)
                                  (Numeric Radius)
                                  (Numeric ODist)
                                  (Numeric Alpha)
                                  (Numeric Theta1)
                                  (Numeric IDist)
                                  (Numeric Theta2)
                                  (Numeric Gamma)
                                  (Angle Alpha)
                                  (Angle Theta1)
                                  (Angle Theta2)
                                  (Angle Gamma)
                                  (Distance Radius)
                                  (Distance ODist)
                                  (Distance IDist)
                                  (SameType (Radius IDist))
                                  (SameType (ODist IDist))
                                  (SameType (Alpha Theta2))
                                  (SameType (Theta1 Theta2))
                                  (SameType (Theta1 Gamma))
                                  (SameType (Alpha Gamma)))
                       (* Type information)
                       (ADD-TO WM (Values Substance glass flint diamond)
                                  (Values Radius plane 100.0 120.0 140.0 -100.0 -120.0 -140.0)
                                  (Values ODist    50.0 -100.0 -150.0 -200.0 -250.0)
                                  (Values Alpha 10.0 15.0 20.0 -10.0 -15.0 -20.0))
                       (* Value information)
                       (ADD-TO WM (CurrentExperiment NIL)
                                  (ExpsInFocus: (exp-1))
                                  (ExpsInNoteBook:)
                                  (Last experiment exp-28)
                                  (FailureCount 0)
                                  (GOAL: run experiment))))
(RPAQQ HM1.CNT (BUILD-IN HM (* * HYPOTHESES MEMORY. It's here where the learned hypotheses rules
                                 will be stored and managed * *)
                            (Predict.PredictionDone ((GOAL: predict))
                                                    -->
                                                    (($DELETE-GOAL WM #1)
                                                     ($CALL CONTROL)))
                            (* Single rules saying that nothing can be predicted)))
```

APPENDIX VI.2

HDD-SH On Three EXPERIMENTS

```
;; Start by setting the goal to design an experiment
CYCLE 1
PRODUCTION: GenDesign2.SetSubGoals
CALL: ACTIVE MEMORIES: (CONTROL)
DELETING GOAL FROM MEMORY WM
(GOAL: run experiment)
ADDING GOAL TO MEMORY WM
(GOAL: design experiment)
-------------------------

CYCLE 2
PRODUCTION: GenDesign0.DesignExp
DELETING GOAL FROM MEMORY WM
(GOAL: design experiment)
ADDING GOAL TO MEMORY WM
(GOAL: get design values)
(GOAL: focus on experiment (exp-1))
-------------------------

CYCLE 3
PRODUCTION: HH.FocusDone
DELETING GOAL FROM MEMORY WM
(GOAL: focus on experiment (exp-1))
-------------------------

CYCLE 4
PRODUCTION: GenDesign0.GetValues

ADD&LINK ENV:
(Exp exp-1 Substance glass)
   ACTIVATION = 1.0glass
ADD&LINK ENV:
(Exp exp-1 Radius plane)
   ACTIVATION = 1.0plane
ADD&LINK ENV:
(Exp exp-1 ODist v:-100.0)
   ACTIVATION = 1.0v:-100.0
ADD&LINK ENV:
(Exp exp-1 Alpha v:10.0)
   ACTIVATION = 1.0v:10.0
ADD&LINK ENV:
(Exp exp-1 Theta1 v:-10.0)
   ACTIVATION = 1.0v:-10.0

DELETING GOAL FROM MEMORYWM
(GOAL: get design values)
ADDING GOAL TO MEMORY WM
(GOAL: predict)

;; The first experiment (#1) is read into ENV (and at the same
time into WM)
-------------------------

CYCLE 5
PRODUCTION: Pred&Comp0.MakePrediction
CALL: ACTIVE MEMORIES: (HM)
ADDING GOAL TO MEMORY WM
(GOAL: evaluate prediction)
-------------------------

CYCLE 6
PRODUCTION: Predict.PredictionDone
DELETING GOAL FROM MEMORYWM
(GOAL: predict)
CALL: ACTIVE MEMORIES: (CONTROL)

;; No prediction could be made since no hypothesis exists yet
-------------------------

CYCLE 7
PRODUCTION: Pred&Comp0.EvalPred
DELETING GOAL FROM MEMORYWM
(GOAL: evaluate prediction)
```

```
ADDING GOAL TO MEMORY WM
(GOAL: compare prediction feedback)
(GOAL: get feedback)
------------------------
CYCLE 8
PRODUCTION: Pred&Comp0.GetFeedbackValues1
ADD&LINK ENV:
(Exp exp-1 Theta2 v:4.0)
  ACTIVATION = 1.0v:4.0
ADD&LINK ENV:
(Exp exp-1 Gamma v:4.0)
  ACTIVATION = 1.0v:4.0
ADD&LINK ENV:
(Exp exp-1 IDist v:-250.0)
  ACTIVATION = 1.0v:-250.0
ADD&LINK ENV:
(Exp exp-1 Feedback W)
  ACTIVATION = 1.0W

DELETING GOAL FROM MEMORY WM
(GOAL: get feedback)

;; The feedback values are read in
------------------------
CYCLE 9
PRODUCTION: Pred&Comp0.NoH
DELETING GOAL FROM MEMORY WM
(GOAL: compare prediction feedback)
ADDING GOAL TO MEMORY WM
(GOAL: generate hypotheses)
------------------------
FIRING 10
PRODUCTION: GenH1.BlockGeneration
DELETING GOAL FROM MEMORY WM
(GOAL: generate hypotheses)
***I removed(GOAL: generate hypotheses) since this is the first experiment
ADDING GOAL TO MEMORY WM
(GOAL: run experiment)
------------------------

CYCLE 11
FIRING 11
PRODUCTION: GenDesign2.SetSubGoals
CALL: ACTIVE MEMORIES: (CONTROL)
DELETING GOAL FROM MEMORY WM
(GOAL: run experiment)
ADDING GOAL TO MEMORY WM
(GOAL: design experiment)

;; The next experiment (#2) is read in
------------------------
CYCLE 12
FIRING 12
PRODUCTION: GenDesign0.DesignExp
DELETING GOAL FROM MEMORY WM
(GOAL: design experiment)
ADDING GOAL TO MEMORY WM
(GOAL: get design values)
(GOAL: focus on experiment (exp-2))
------------------------
CYCLE 15
FIRING 15
PRODUCTION: GenDesign0.GetValues
ADD&LINK ENV:
(Exp exp-2 Substance glass)
  ACTIVATION = 1.0glass
ADD&LINK ENV:
(Exp exp-2 Radius plane)
  ACTIVATION = 1.0plane
ADD&LINK ENV:
(Exp exp-2 ODist v:-100.0)
  ACTIVATION = 1.0v:-100.0
ADD&LINK ENV:
(Exp exp-2 Alpha v:15.0)
  ACTIVATION = 1.0v:15.0
```

```
ADD&LINK ENV:
(Exp exp-2 Theta1 v:-15.0)
  ACTIVATION = 1.0v:-15.0

DELETING GOAL FROM MEMORY WM
(GOAL: get design values)
ADDING GOAL TO MEMORY WM
(GOAL: predict)
-------------------------
CYCLE 16
FIRING 16
PRODUCTION: Pred&Comp0.MakePrediction
CALL: ACTIVE MEMORIES: (HM)
ADDING GOAL TO MEMORY WM
(GOAL: evaluate prediction)
-------------------------
CYCLE 17
FIRING 17
PRODUCTION: Predict.PredictionDone
DELETING GOAL FROM MEMORYWM
(GOAL: predict)
CALL: ACTIVE MEMORIES: (CONTROL)
```

;; Again, no prediction possible

```
-------------------------
CYCLE 18
FIRING 18
PRODUCTION: Pred&Comp0.EvalPred
DELETING GOAL FROM MEMORY WM
(GOAL: evaluate prediction)
ADDING GOAL TO MEMORY WM
(GOAL: compare prediction feedback)
(GOAL: get feedback)
-------------------------
CYCLE 19
FIRING 19
PRODUCTION: Pred&Comp0.GetFeedbackValues
ADD&LINK ENV:
(Exp exp-2 Theta2 v:6.0)
  ACTIVATION = 1.0v:6.0
ADD&LINK ENV:
(Exp exp-2 Gamma v:6.0)
  ACTIVATION = 1.0v:6.0
ADD&LINK ENV:
(Exp exp-2 IDist v:-250.0)
  ACTIVATION = 1.0v:-250.0
ADD&LINK ENV:
(Exp exp-2 Feedback W)
  ACTIVATION = 1.0W
DELETING GOAL FROM MEMORYWM
(GOAL: get feedback)
-------------------------
CYCLE 20
FIRING 20
PRODUCTION: Pred&Comp0.NoH
DELETING GOAL FROM MEMORYWM
(GOAL: compare prediction feedback)
ADDING GOAL TO MEMORY WM
(GOAL: generate hypotheses)
```

;; The system begins now to generate first hypotheses since it has data from two experiments available. Data from the last experiment (#1) are read in.

```
-------------------------
CYCLE 21
FIRING 21
PRODUCTION: GenH1.ShiftFocus
ADDING GOAL TO MEMORY WM
(GOAL: consider variables)
(GOAL: focus on experiment (exp 1 exp-2))
-------------------------
CYCLE 22
FIRING 22
PRODUCTION: HH.Focus
**RemovingNILfromfocus

**Moving(exp-1)intofocus
```

```
** Searching for information about exp-1
   in ENV...
-------------------------
;; Now the trend detectors can apply

CYCLE 26
FIRING 32
PRODUCTION: GenH1.CallTrendDetector2
ADDING GOAL TO MEMORY WM
(GOAL: find trends)
-------------------------
CYCLE 27
FIRING 33
PRODUCTION: TREND.Done
CALL: ACTIVE MEMORIES: (CONTROL)
FIRING 34
PRODUCTION: TREND.Direct1
***AsAlphaincreases,Theta2increases

FIRING 35
PRODUCTION: TREND.Direct2
***AsAlphadecreasesTheta2decreases

FIRING 36
PRODUCTION: TREND.Inverse1
***AsTheta1increasesTheta2decreases

FIRING 37
PRODUCTION: TREND.Inverse2
***AsTheta1decreasesTheta2increases

FIRING 38
PRODUCTION: TREND.Direct1
***AsAlphaincreases,Gammaincreases

FIRING 39
PRODUCTION: TREND.Direct2
***AsAlphadecreasesGammadecreases

FIRING 40
PRODUCTION: TREND.Inverse1
***AsTheta1increasesGammadecreases

FIRING 41
PRODUCTION: TREND.Inverse2
***AsTheta1decreasesGammaincreases
-------------------------
CYCLE 28
FIRING 42
PRODUCTION: GenH1.TrendDetectorDone
DELETING GOAL FROM MEMORY WM
(GOAL: find trends)
ADDING GOAL TO MEMORY WM
(GOAL: focus on experiment (exp-2))

;; Experiment #1 is no longer needed and removed from WM. Then
   function induction starts.
-------------------------
(...)

CYCLE 34
FIRING 57
PRODUCTION: FUNCTION.Quotient
***Ifound:(Function: DV: Gamma IV: Alpha Form: (
LAMBDA (iv) (ROUNDED.QUOTIENT iv 2.5)))
BUILDING   Hypothesis-003

FIRING 58
PRODUCTION: FUNCTION.Difference
***Ifound:(Function: DV: Gamma IV: Alpha Form: (
LAMBDA (IV) (ROUNDED.DIFFERENCE IV 9.0)))
BUILDING   Hypothesis-0032
```

FIRING 59
PRODUCTION: FUNCTION.Quotient
***Ifound:(Function: DV: Theta2 IV: Alpha Form:
(LAMBDA (iv) (ROUNDED.QUOTIENT iv 2.5)))
BUILDING   Hypothesis-0033

FIRING 60
PRODUCTION: FUNCTION.Difference
***Ifound:(Function: DV: Theta2 IV: Alpha Form:
(LAMBDA (IV) (ROUNDED.DIFFERENCE IV 9.0)))
BUILDING   Hypothesis-0034

FIRING 61
PRODUCTION: FUNCTION.Product
***Ifound:(Function: DV: Gamma IV: Theta1 Form:
(LAMBDA (iv) (ROUNDED.TIMES iv -.4)))
BUILDING   Hypothesis-0035

FIRING 62
PRODUCTION: FUNCTION.Sum
***Ifound:(Function: DV: Gamma IV: Theta1 Form:
(LAMBDA (IV) (ROUNDED.PLUS IV 21.0)))
BUILDING   Hypothesis-0036

FIRING 63
PRODUCTION: FUNCTION.Product
***Ifound:(Function: DV: Theta2 IV: Theta1 Form:
 (LAMBDA (iv) (ROUNDED.TIMES iv -.4)))
BUILDING   Hypothesis-0037

FIRING 64
PRODUCTION: FUNCTION.Sum
***Ifound:(Function: DV: Theta2 IV: Theta1 Form:
 (LAMBDA (IV) (ROUNDED.PLUS IV 21.0)))
BUILDING   Hypothesis-0038

;; For each functional relation found, a new rule was created
and stored in HMEM Experiment number #3 starts.

CYCLE 35
FIRING 65
PRODUCTION: GenH1.Done1
DELETING GOAL FROM MEMORYWM
(GOAL: find functions)
ADDING GOAL TO MEMORY WM
(GOAL: run experiment)

CYCLE 36
FIRING 66
PRODUCTION: GenDesign2.SetSubGoals
CALL: ACTIVE MEMORIES: (CONTROL)
DELETING GOAL FROM MEMORYWM
(GOAL: run experiment)
ADDING GOAL TO MEMORY WM
(GOAL: design experiment)

CYCLE 37
FIRING 67
PRODUCTION: GenDesign0.DesignExp
DELETING GOAL FROM MEMORYWM
(GOAL: design experiment)
ADDING GOAL TO MEMORY WM
(GOAL: get design values)
(GOAL: focus on experiment (exp-3))

CYCLE 40
FIRING 70
PRODUCTION: GenDesign0.GetValues
ADD&LINK ENV:
(Exp exp-3 Substance glass)
   ACTIVATION = 1.0glass
ADD&LINK ENV:
(Exp exp-3 Radius plane)
   ACTIVATION = 1.0plane
ADD&LINK ENV:
(Exp exp-3 ODist v:-200.0)
   ACTIVATION = 1.0v:-200.0
ADD&LINK ENV:

```
(Exp exp-3 Alpha v:10.0)
   ACTIVATION = 1.0v:10.0
ADD&LINK ENV:
(Exp exp-3 Theta1 v:-10.0)
   ACTIVATION = 1.0v:-10.0
DELETING GOAL FROM MEMORY WM
(GOAL: get design values)
ADDING GOAL TO MEMORY WM
(GOAL: predict)
-------------------------
CYCLE 41
FIRING 71
PRODUCTION: Pred&Comp0.MakePrediction
CALL: ACTIVE MEMORIES: (HM)
ADDING GOAL TO MEMORY WM
(GOAL: evaluate prediction)
-------------------------
CYCLE 42
FIRING 72
PRODUCTION: Hypothesis-0032
***Myhypothesisis:Gamma equals Alpha(LAMBDA (IV) (
ROUNDED.DIFFERENCE IV 9.0))

CALL: ACTIVE MEMORIES: (CONTROL)
```

;; This time, a prediction is delivered based on
Hypothesis-0032. Next, the prediction is compared with the
feedback.

```
-------------------------
CYCLE 43
FIRING 73
PRODUCTION: Pred&Comp0.EvalPred
DELETING GOAL FROM MEMORYWM
(GOAL: evaluate prediction)
ADDING GOAL TO MEMORY WM
(GOAL: compare prediction feedback)
(GOAL: get feedback)
-------------------------
CYCLE 44
FIRING 74
PRODUCTION: Pred&Comp0.GetFeedbackValue
ADD&LINK ENV:
(Exp exp-3 Theta2 v:4.0)
   ACTIVATION = 1.0v:4.0
ADD&LINK ENV:
(Exp exp-3 Gamma v:4.0)
   ACTIVATION = 1.0v:4.0
ADD&LINK ENV:
(Exp exp-3 IDist v:-500.0)
   ACTIVATION = 1.0v:-500.0
ADD&LINK ENV:
(Exp exp-3 Feedback W)
   ACTIVATION = 1.0W
DELETING GOAL FROM MEMORYWM
(GOAL: get feedback)
-------------------------
CYCLE 45
FIRING 75
PRODUCTION: Pred&Comp0.Compare1
ADDING GOAL TO MEMORY WM
(GOAL: compare (Prediction-0039 ISA:
Prediction DV: Gamma IV: Alpha Form: (LAMBDA
(IV) (ROUNDED.DIFFERENCE IV 9.0)) Value: 1.0)
exp-3)
```

;; It turns out to be wrong and the discrimination process is
triggered

```
-------------------------
CYCLE 46
FIRING 76
PRODUCTION: Pred&Comp1.WrongPred
DELETING GOAL FROM MEMORYWM
(GOAL: compare (Prediction-0039 ISA:
Prediction DV: Gamma IV: Alpha Form: (LAMBDA
(IV) (ROUNDED.DIFFERENCE IV 9.0)) Value: 1.0)
exp-3)
```

***Predictioniswrong

CYCLE 47
FIRING 77
PRODUCTION: Pred&Comp0.CompareDone
DELETING GOAL FROM MEMORYWM
(GOAL: compare prediction feedback)
ADDING GOAL TO MEMORY WM
(GOAL: eval hypotheses)

CYCLE 48
FIRING 78
PRODUCTION: ModifyH2.WrongPred
***Predictioniswrong...Itrytofindnewconditions
[48] (Prediction-0039 ISA: Prediction DV: Gamma
IV: Alpha Form: (LAMBDA (IV) (ROUNDED.DIFFERENCE
 IV 9.0)) Value: 1.0)based on hypothesis
Hypothesis-0032 is wrong

The new attribute value for the rule
Hypothesis-0032 is: .25

;; **The strength of the wrong hypothesis is decreased and new, more specific variants are created**

Discriminated rule: Hypothesis-0032-1
((GOAL: predict)
 (Exp =e Substance =iv1)
 (Exp =e Radius =iv2)
 (Exp =e ODist =iv3)
 (Exp =e Alpha =iv4)
 (Exp =e Theta1 -15.0))

Discriminated rule: Hypothesis-0032-2
((GOAL: predict)
 (Exp =e Substance =iv1)
 (Exp =e Radius =iv2)
 (Exp =e ODist -100.0)
 (Exp =e Alpha =iv4)
 (Exp =e Theta1 =iv6))

HM-VANISH EVOKED:
UNBUILDINGHypothesis-0032
[48] Transfer process HM-VANISH removed
Hypothesis-0032

;; **H-0032 is forgotten and the hypothesis generatot is called to look for new equations.**

CYCLE 49
FIRING 79
PRODUCTION: ModifyH0.CallGenerator
DELETING GOAL FROM MEMORYWM
(GOAL: eval hypotheses)
(A hypothesis was wrong)
ADDING GOAL TO MEMORY WM
(GOAL: generate hypotheses)

CYCLE 50
FIRING 80
PRODUCTION: GenH1.ShiftFocus
ADDING GOAL TO MEMORY WM
(GOAL: consider variables)
(GOAL: focus on experiment (exp-2 exp-3))

(...)
CYCLE 53
FIRING 83
PRODUCTION: GenH1.CallTrendDetector1

(...)

APPENDIX VI.3

HDD-SH on 27 Experiments (Product Trace)'

Experiment exp-1
[4]Design: glass plane v:-100.0 v:10.0 v:-10.0
[8]Feedback: v:4.0 v:4.0 v:-250.0 W

Experiment exp-2
[15]Design: glass plane v:-100.0 v:15.0 v:-15.0
[19]Feedback: v:6.0 v:6.0 v:-250.0 W
[34] Attribute SIMPLICITYof Hypothesis-0031 changed: (1.0 12 3.0)
[34] Rule newly built:
Hypothesis-0031 (Hypothesis: DV: Gamma IV: Alpha Form: (LAMBDA (iv) (ROUNDED.QUOTIENT iv 2.5)))
[34] Attribute SIMPLICITYof Hypothesis-0032 changed: (1.0 12 3.0)
[34] Rule newly built:
Hypothesis-0032 (Hypothesis: DV: Gamma IV: Alpha Form: (LAMBDA (IV) (ROUNDED.DIFFERENCE IV 9.0)))
[34] Attribute SIMPLICITYof Hypothesis-0033 changed: (1.0 12 3.0)
[34] Rule newly built:
Hypothesis-0033 (Hypothesis: DV: Theta2 IV: Alpha Form: (LAMBDA (iv) (ROUNDED.QUOTIENT iv 2.5)))
[34] Attribute SIMPLICITYof Hypothesis-0034 changed: (1.0 12 3.0)
[34] Rule newly built:
Hypothesis-0034 (Hypothesis: DV: Theta2 IV: Alpha Form: (LAMBDA (IV) (ROUNDED.DIFFERENCE IV 9.0)))
[34] Attribute SIMPLICITYof Hypothesis-0035 changed: (1.0 12 2.0)
[34] Rule newly built:
Hypothesis-0035 (Hypothesis: DV: Gamma IV: Theta1 Form: (LAMBDA (iv) (ROUNDED.TIMES iv -.4)))
[34] Attribute SIMPLICITYof Hypothesis-0036 changed: (1.0 12 3.0)
[34] Rule newly built:
Hypothesis-0036 (Hypothesis: DV: Gamma IV: Theta1 Form: (LAMBDA (IV) (ROUNDED.PLUS IV 21.0)))
[34] Attribute SIMPLICITYof Hypothesis-0037 changed: (1.0 12 2.0)
[34] Rule newly built:
Hypothesis-0037 (Hypothesis: DV: Theta2 IV: Theta1 Form: (LAMBDA (iv) (ROUNDED.TIMES iv -.4)))
[34] Attribute SIMPLICITYof Hypothesis-0038 changed: (1.0 12 3.0)
[34] Rule newly built:
Hypothesis-0038 (Hypothesis: DV: Theta2 IV: Theta1 Form: (LAMBDA (IV) (ROUNDED.PLUS IV 21.0)))

Experiment exp-3
[40]Design: glass plane v: 200.0 v:10.0 v:-10.0
[44]Feedback: v:4.0 v:4.0 v:-500.0 W
[48] (Prediction-0039 ISA: Prediction DV: Theta2 IV: Theta1 Form: (LAMBDA (iv) (ROUNDED.TIMES iv -.4))
Value: 4.0)based on hypothesis Hypothesis-0037 is correct
[48] The new attribute value for the rule Hypothesis-0037 is: 2.0

Experiment exp-4
[54]Design: glass plane v:-200.0 v:15.0 v:-15.0
[58]Feedback: v:6.0 v:6.0 v:-500.0 W
[62] (Prediction-0040 ISA: Prediction DV: Theta2 IV: Theta1 Form: (LAMBDA (iv) (ROUNDED.TIMES iv -.4))
Value: 6.0)based on hypothesis Hypothesis-0037 is correct
[62] The new attribute value for the rule Hypothesis-0037 is: 3.0

Experiment exp-5
[68]Design: glass v:100.0 v: 100.0 v:10.0 v: 20.5
[72]Feedback: v:8.2 v:-2.1 v:500.0 W
[76] (Prediction-0041 ISA: Prediction DV: Theta2 IV: Theta1 Form: (LAMBDA (iv) (ROUNDED.TIMES iv -.4))
Value: 8.2)based on hypothesis Hypothesis-0037 is correct
[76] The new attribute value for the rule Hypothesis-0037 is: 4.0

Experiment exp-6
[82]Design: glass v:100.0 v:-100.0 v:15.0 v:-31.4
[86]Feedback: v:12.5 v:-3.2 v:500.0 W
[90] (Prediction-0042 ISA: Prediction DV: Theta2 IV: Theta1 Form: (LAMBDA (iv) (ROUNDED.TIMES iv -.4))
Value: 12.56)based on hypothesis Hypothesis-0037 is correct
[90] The new attribute value for the rule Hypothesis-0037 is: 5.0

Experiment exp-7
[96]Design: glass v:100.0 v: 200.0 v:10.0 v: 32.5
[100]Feedback: v:12.9 v: 8.8 v:250.0 W
[104] (Prediction-0043 ISA: Prediction DV: Theta2 IV: Theta1 Form: (LAMBDA (iv) (ROUNDED.TIMES iv -.4))

) Value: 12.92)based on hypothesis Hypothesis-0037 is correct
[104] The new attribute value for the rule Hypothesis-0037 is: 6.0

Experiment exp-8
[110] Design: glass v:100.0 v:-200.0 v:15.0 v:-50.7
[114] Feedback: v:20.3 v:-14.2 v:250.0 W
[118] (Prediction-0044 ISA: Prediction DV: Theta2 IV: Theta1 Form: (LAMBDA (iv) (ROUNDED.TIMES iv -.4)
) Value: 20.28)based on hypothesis Hypothesis-0037 is correct
[118] The new attribute value for the rule Hypothesis-0037 is: 7.0

Experiment exp-9
[124] Design: flint plane v:-100.0 v:10.0 v:-10.0
[128] Feedback: v:3.3 v:3.3 v:-300.0 W
[132] (Prediction-0045 ISA: Prediction DV: Theta2 IV: Theta1 Form: (LAMBDA (iv) (ROUNDED.TIMES iv -.4)
) Value: 4.0)based on hypothesis Hypothesis-0037 is wrong
[132] The new attribute value for the rule Hypothesis-0037 is: 1.75
[132] Attribute SIMPLICITYof Hypothesis-0037-1 changed: (1.0 13 2.0)
[132] Discriminated rule: Hypothesis-0037-1
((GOAL: predict)
 (Exp =e Substance =iv1)
 (Exp =e Radius =iv2)
 (Exp =e ODist =iv3)
 (Exp =e Alpha 15.0)
 (Exp =e Theta1 =iv6))
[132] Attribute SIMPLICITYof Hypothesis-0037-2 changed: (1.0 13 2.0)
[132] Discriminated rule: Hypothesis-0037-2
((GOAL: predict)
 (Exp =e Substance =iv1)
 (Exp =e Radius =iv2)
 (Exp =e ODist -200.0)
 (Exp =e Alpha =iv4)
 (Exp =e Theta1 =iv6))
[132] Attribute SIMPLICITYof Hypothesis-0037-3 changed: (1.0 13 2.0)
[132] Discriminated rule: Hypothesis-0037-3
((GOAL: predict)
 (Exp =e Substance =iv1)
 (Exp =e Radius 100.0)
 (Exp =e ODist =iv3)
 (Exp =e Alpha =iv4)
 (Exp =e Theta1 =iv6))
[132] Attribute SIMPLICITYof Hypothesis-0037-4 changed: (1.0 13 2.0)
[132] Discriminated rule: Hypothesis-0037-4
((GOAL: predict)
 (Exp =e Substance glass)
 (Exp =e Radius =iv2)
 (Exp =e ODist =iv3)
 (Exp =e Alpha =iv4)
 (Exp =e Theta1 =iv6))
[147] Attribute SIMPLICITYof Hypothesis-0046 changed: (1.0 12 1.0)
[147] Rule newly built:
Hypothesis-0046 (Hypothesis: DV: Theta2 IV: Alpha Form: (LAMBDA (iv) (ROUNDED.QUOTIENT iv 3.03)))
[147] Attribute SIMPLICITYof Hypothesis-0047 changed: (1.0 12 2.0)
[147] Rule newly built:
Hypothesis-0047 (Hypothesis: DV: Theta2 IV: Alpha Form: (LAMBDA (IV) (ROUNDED.DIFFERENCE IV 6.7)))
[147] Attribute SIMPLICITYof Hypothesis-0048 changed: (1.0 12 1.0)
[147] Rule newly built:
Hypothesis-0048 (Hypothesis: DV: Gamma IV: Theta1 Form: (LAMBDA (iv) (ROUNDED.QUOTIENT iv -3.03)))
[147] Attribute SIMPLICITYof Hypothesis-0049 changed: (1.0 12 2.0)
[147] Rule newly built:
Hypothesis-0049 (Hypothesis: DV: Gamma IV: Theta1 Form: (LAMBDA (IV) (ROUNDED.DIFFERENCE IV -13.3)))
[147] Attribute SIMPLICITYof Hypothesis-0050 changed: (1.0 12 3.0)
[147] Rule newly built:
Hypothesis-0050 (Hypothesis: DV: IDist IV: ODist Form: (LAMBDA (iv) (ROUNDED.TIMES iv 3.0)))
[147] Attribute SIMPLICITYof Hypothesis-0051 changed: (1.0 12 3.0)
[147] Rule newly built:
Hypothesis-0051 (Hypothesis: DV: IDist IV: ODist Form: (LAMBDA (IV) (ROUNDED.PLUS IV -200.0)))
[147] Attribute SIMPLICITYof Hypothesis-0052 changed: (1.0 12 1.0)
[147] Rule newly built:
Hypothesis-0052 (Hypothesis: DV: Theta2 IV: Theta1 Form: (LAMBDA (iv) (ROUNDED.TIMES iv -.33)))
[147] Attribute SIMPLICITYof Hypothesis-0053 changed: (1.0 12 2.0)
[147] Rule newly built:
Hypothesis-0053 (Hypothesis: DV: Theta2 IV: Theta1 Form: (LAMBDA (IV) (ROUNDED.PLUS IV 13.3)))

[147] Attribute SIMPLICITYof Hypothesis-0054 changed: (1.0 12 1.0)
[147] Rule newly built:
Hypothesis-0054 (Hypothesis: DV: Gamma IV: Alpha Form: (LAMBDA (iv) (ROUNDED.TIMES iv .33)))
[147] Attribute SIMPLICITYof Hypothesis-0055 changed: (1.0 12 2.0)
[147] Rule newly built:
Hypothesis-0055 (Hypothesis: DV: Gamma IV: Alpha Form: (LAMBDA (IV) (ROUNDED.PLUS IV -6.7)))

Experiment exp-10
[153]Design: flint plane v:-100.0 v:15.0 v:-15.0
[157]Feedback: v:5.0 v:5.0 v:-300.0 W
[161] (Prediction-0056 ISA: Prediction DV: Theta2 IV: Theta1 Form: (LAMBDA (iv) (ROUNDED.TIMES iv -.4)
) Value: 6.0)based on hypothesis Hypothesis-0037 is wrong
[161] The new attribute value for the rule Hypothesis-0037 is: .4375
[161] The new attribute value for the rule Hypothesis-0037-1 is: 2.0
[161] The new attribute value for the rule Hypothesis-0037-2 is: 2.0
[161] The new attribute value for the rule Hypothesis-0037-3 is: 2.0
[161] The new attribute value for the rule Hypothesis-0037-4 is: 2.0
[161] Transfer process HM-VANISH removed Hypothesis-0037
[176] Attribute SIMPLICITYof Hypothesis-0057 changed: (1.0 12 3.0)
[176] Rule newly built:
Hypothesis-0057 (Hypothesis: DV: Theta2 IV: Alpha Form: (LAMBDA (iv) (ROUNDED.QUOTIENT iv 3.0)))
[176] Attribute SIMPLICITYof Hypothesis-0058 changed: (1.0 12 3.0)
[176] Rule newly built:
Hypothesis-0058 (Hypothesis: DV: Theta2 IV: Alpha Form: (LAMBDA (IV) (ROUNDED.DIFFERENCE IV 10.0)))
[176] Attribute SIMPLICITYof Hypothesis-0059 changed: (1.0 12 3.0)
[176] Rule newly built:
Hypothesis-0059 (Hypothesis: DV: Gamma IV: Alpha Form: (LAMBDA (iv) (ROUNDED.QUOTIENT iv 3.0)))
[176] Attribute SIMPLICITYof Hypothesis-0060 changed: (1.0 12 3.0)
[176] Rule newly built:
Hypothesis-0060 (Hypothesis: DV: Gamma IV: Alpha Form: (LAMBDA (IV) (ROUNDED.DIFFERENCE IV 10.0)))
[176] Rule: Hypothesis-0052 rebuilt.
[176] Attribute SIMPLICITYof Hypothesis-0062 changed: (1.0 12 3.0)
[176] Rule newly built:
Hypothesis-0062 (Hypothesis: DV: Theta2 IV: Theta1 Form: (LAMBDA (IV) (ROUNDED.PLUS IV 20.0)))
[176] Attribute SIMPLICITYof Hypothesis-0063 changed: (1.0 12 1.0)
[176] Rule newly built:
Hypothesis-0063 (Hypothesis: DV: Gamma IV: Theta1 Form: (LAMBDA (iv) (ROUNDED.TIMES iv -.33)))
[176] Attribute SIMPLICITYof Hypothesis-0064 changed: (1.0 12 3.0)
[176] Rule newly built:
Hypothesis-0064 (Hypothesis: DV: Gamma IV: Theta1 Form: (LAMBDA (IV) (ROUNDED.PLUS IV 20.0)))

Experiment exp-11
[182]Design: flint plane v:-200.0 v:10.0 v:-10.0
[186]Feedback: v:3.3 v:3.3 v:-600.0 W
[190] (Prediction-0065 ISA: Prediction DV: Theta2 IV: Theta1 Form: (LAMBDA (iv) (ROUNDED.TIMES iv -.33
)) Value: 3.3)based on hypothesis Hypothesis-0052 is correct
[190] The new attribute value for the rule Hypothesis-0052 is: 3.0

Experiment exp-12
[196]Design: flint plane v:-200.0 v:15.0 v:-15.0
[200]Feedback: v:5.0 v:5.0 v:-600.0 W
[204] (Prediction-0066 ISA: Prediction DV: Theta2 IV: Theta1 Form: (LAMBDA (iv) (ROUNDED.TIMES iv -.33
)) Value: 4.95)based on hypothesis Hypothesis-0052 is correct
[204] The new attribute value for the rule Hypothesis-0052 is: 4.0

Experiment exp-13
[210]Design: flint v:100.0 v:-100.0 v:10.0 v:-20.5
[214]Feedback: v:6.8 v:-3.5 v:300.0 W
[218] (Prediction-0067 ISA: Prediction DV: Theta2 IV: Theta1 Form: (LAMBDA (iv) (ROUNDED.TIMES iv -.33
)) Value: 6.77)based on hypothesis Hypothesis-0052 is correct
[218] The new attribute value for the rule Hypothesis-0052 is: 5.0

Experiment exp-14
[224]Design: flint v:100.0 v:-100.0 v:15.0 v:-31.4
[228]Feedback: v:10.4 v:-5.4 v:300.0 W
[232] (Prediction-0068 ISA: Prediction DV: Theta2 IV: Theta1 Form: (LAMBDA (iv) (ROUNDED.TIMES iv -.33
)) Value: 10.36)based on hypothesis Hypothesis-0052 is correct

[232] The new attribute value for the rule Hypothesis-0052 is: 6.0

Experiment exp-15
[238] Design: flint v:100.0 v:-200.0 v:10.0 v: 32.3
[242] Feedback: v:10.7 v:-11.0 v:200.0 W
[246] (Prediction 0069 ISA: Prediction DV: Theta2 IV: Theta1 Form: (LAMBDA (iv) (ROUNDED.TIMES iv -.33
)) Value: 10.66)based on hypothesis Hypothesis-0052 is correct
[246] The new attribute value for the rule Hypothesis-0052 is: 7.0

Experiment exp-16
[252] Design: flint v:100.0 v:-200.0 v:15.0 v:-50.7
[256] Feedback: v:16.8 v:-17.9 v:200.0 W
[260] (Prediction 0070 ISA: Prediction DV: Theta2 IV: Theta1 Form: (LAMBDA (iv) (ROUNDED.TIMES iv -.33
)) Value: 16.73)based on hypothesis Hypothesis-0052 is correct
[260] The new attribute value for the rule Hypothesis-0052 is: 8.0

Experiment exp-17
[266] Design: diamond plane v:-100.0 v:10.0 v:-10.0
[270] Feedback: v:2.8 v:2.8 v:-350.0 W
[274] (Prediction 0071 ISA: Prediction DV: Theta2 IV: Theta1 Form: (LAMBDA (iv) (ROUNDED.TIMES iv -.33
)) Value: 3.3)based on hypothesis Hypothesis-0052 is wrong
[274] The new attribute value for the rule Hypothesis-0052 is: 2.0
[274] Attribute SIMPLICITYof Hypothesis-0052-1 changed: (1.0 13 1.0)
[274] Discriminated rule: Hypothesis-0052-1
((GOAL: predict)
 (Exp =e Substance =iv1)
 (Exp =e Radius =iv2)
 (Exp =e ODist =iv3)
 (Exp =e Alpha 15.0)
 (Exp =e Theta1 =iv6))
[274] Attribute SIMPLICITYof Hypothesis-0052-2 changed: (1.0 13 1.0)
[274] Discriminated rule: Hypothesis-0052-2
((GOAL: predict)
 (Exp =e Substance =iv1)
 (Exp =e Radius =iv2)
 (Exp =e ODist -200.0)
 (Exp =e Alpha =iv4)
 (Exp =e Theta1 =iv6))
[274] Attribute SIMPLICITYof Hypothesis-0052-3 changed: (1.0 13 1.0)
[274] Discriminated rule: Hypothesis-0052-3
((GOAL: predict)
 (Exp =e Substance =iv1)
 (Exp =e Radius 100.0)
 (Exp =e ODist =iv3)
 (Exp =e Alpha =iv4)
 (Exp =e Theta1 =iv6))
[274] Attribute SIMPLICITYof Hypothesis-0052-4 changed: (1.0 13 1.0)
[274] Discriminated rule: Hypothesis-0052-4
((GOAL: predict)
 (Exp =e Substance flint)
 (Exp =e Radius =iv2)
 (Exp =e ODist =iv3)
 (Exp =e Alpha =iv4)
 (Exp =e Theta1 =iv6))
[289] Attribute SIMPLICITYof Hypothesis-0072 changed: (1.0 12 1.0)
[289] Rule newly built:
Hypothesis 0072 (Hypothesis: DV: Gamma IV: Theta1 Form: (LAMBDA (iv) (ROUNDED.QUOTIENT iv -3.57)))
[289] Attribute SIMPLICITYof Hypothesis-0073 changed: (1.0 12 2.0)
[289] Rule newly built:
Hypothesis 0073 (Hypothesis: DV: Gamma IV: Theta1 Form: (LAMBDA (IV) (ROUNDED.DIFFERENCE IV -12.8)))
[289] Attribute SIMPLICITYof Hypothesis-0074 changed: (1.0 12 1.0)
[289] Rule newly built:
Hypothesis 0074 (Hypothesis: DV: Theta2 IV: Alpha Form: (LAMBDA (iv) (ROUNDED.QUOTIENT iv 3.57)))
[289] Attribute SIMPLICITYof Hypothesis-0075 changed: (1.0 12 2.0)
[289] Rule newly built:
Hypothesis 0075 (Hypothesis: DV: Theta2 IV: Alpha Form: (LAMBDA (IV) (ROUNDED.DIFFERENCE IV 7.2)))
[289] Attribute SIMPLICITYof Hypothesis-0076 changed: (1.0 12 1.0)
[289] Rule newly built:
Hypothesis 0076 (Hypothesis: DV: Gamma IV: Alpha Form: (LAMBDA (iv) (ROUNDED.TIMES iv .28)))
[289] Attribute SIMPLICITYof Hypothesis-0077 changed: (1.0 12 2.0)

```
[289] Rule newly built:
Hypothesis-0077 (Hypothesis: DV: Gamma IV: Alpha Form: (LAMBDA (IV) (ROUNDED.PLUS IV -7.2)))
[289] Attribute SIMPLICITYof Hypothesis-0078 changed: (1.0 12 1.0)
[289] Rule newly built:
Hypothesis-0078 (Hypothesis: DV: Theta2 IV: Theta1 Form: (LAMBDA (iv) (ROUNDED.TIMES iv -.28)))
[289] Attribute SIMPLICITYof Hypothesis-0079 changed: (1.0 12 2.0)
[289] Rule newly built:
Hypothesis-0079 (Hypothesis: DV: Theta2 IV: Theta1 Form: (LAMBDA (IV) (ROUNDED.PLUS IV 12.8)))
[289] Attribute SIMPLICITYof Hypothesis-0080 changed: (1.0 12 3.0)
[289] Rule newly built:
Hypothesis-0080 (Hypothesis: DV: IDist IV: ODist Form: (LAMBDA (iv) (ROUNDED.TIMES iv 3.5)))
[289] Attribute SIMPLICITYof Hypothesis-0081 changed: (1.0 12 3.0)
[289] Rule newly built:
Hypothesis-0081 (Hypothesis: DV: IDist IV: ODist Form: (LAMBDA (IV) (ROUNDED.PLUS IV -250.0)))
```

Experiment exp-18
```
[295]Design: diamond plane v:-100.0 v:15.0 v:-15.0
[299]Feedback: v:4.2 v:4.2 v:-350.0 W
[303] (Prediction 0082 ISA: Prediction DV: Theta2 IV: Theta1 Form: (LAMBDA (iv) (ROUNDED.TIMES iv
-.33
)) Value: 4.95)based on hypothesis Hypothesis-0052 is wrong
[303] The new attribute value for the rule Hypothesis-0052 is: .5
[303] The new attribute value for the rule Hypothesis-0052-1 is: 2.0
[303] The new attribute value for the rule Hypothesis-0052-2 is: 2.0
[303] The new attribute value for the rule Hypothesis-0052-3 is: 2.0
[303] The new attribute value for the rule Hypothesis-0052-4 is: 2.0
[318] Attribute SIMPLICITYof Hypothesis-0083 changed: (1.0 12 1.0)
[318] Rule newly built:
Hypothesis-0083 (Hypothesis: DV: Gamma IV: Alpha Form: (LAMBDA (iv) (ROUNDED.QUOTIENT iv 3.57)))
[318] Attribute SIMPLICITYof Hypothesis-0084 changed: (1.0 12 2.0)
[318] Rule newly built:
Hypothesis-0084 (Hypothesis: DV: Gamma IV: Alpha Form: (LAMBDA (IV) (ROUNDED.DIFFERENCE IV
10.8)))
[318] Rule: Hypothesis-0074 rebuilt.
[318] Attribute SIMPLICITYof Hypothesis-0086 changed: (1.0 12 2.0)
[318] Rule newly built:
Hypothesis-0086 (Hypothesis: DV: Theta2 IV: Alpha Form: (LAMBDA (IV) (ROUNDED.DIFFERENCE IV
10.8)))
[318] Attribute SIMPLICITYof Hypothesis-0087 changed: (1.0 12 1.0)
[318] Rule newly built:
Hypothesis-0087 (Hypothesis: DV: Gamma IV: Theta1 Form: (LAMBDA (iv) (ROUNDED.TIMES iv -.28)))
[318] Attribute SIMPLICITYof Hypothesis-0088 changed: (1.0 12 2.0)
[318] Rule newly built:
Hypothesis-0088 (Hypothesis: DV: Gamma IV: Theta1 Form: (LAMBDA (IV) (ROUNDED.PLUS IV 19.2)))
[318] Rule: Hypothesis-0078 rebuilt.
[318] Attribute SIMPLICITYof Hypothesis-0090 changed: (1.0 12 2.0)
[318] Rule newly built:
Hypothesis-0090 (Hypothesis: DV: Theta2 IV: Theta1 Form: (LAMBDA (IV) (ROUNDED.PLUS IV 19.2)))
```

Experiment exp-19
```
[324]Design: diamond plane v:-200.0 v:10.0 v:-10.0
[328]Feedback: v:2.8 v:2.8 v:-700.0 W
[332] (Prediction-0091 ISA: Prediction DV: Theta2 IV: Alpha Form: (LAMBDA (iv) (ROUNDED.QUOTIENT
iv
3.57)) Value: 2.8)based on hypothesis Hypothesis-0074 is correct
[332] The new attribute value for the rule Hypothesis-0074 is: 3.0
```

Experiment exp-20
```
[338]Design: diamond plane v:-200.0 v:15.0 v:-15.0
[342]Feedback: v:4.2 v:4.2 v:-700.0 W
[346] (Prediction-0092 ISA: Prediction DV: Theta2 IV: Alpha Form: (LAMBDA (iv) (ROUNDED.QUOTIENT
iv
3.57)) Value: 4.2)based on hypothesis Hypothesis-0074 is correct
[346] The new attribute value for the rule Hypothesis-0074 is: 4.0
```

Experiment exp-21
```
[352]Design: diamond v:100.0 v:-100.0 v:10.0 v:-20.5
[356]Feedback: v:5.7 v:-4.5 v:233.0 W
[360] (Prediction-0093 ISA: Prediction DV: Theta2 IV: Alpha Form: (LAMBDA (iv) (ROUNDED.QUOTIENT
iv
3.57)) Value: 2.8)based on hypothesis Hypothesis-0074 is wrong
[360] The new attribute value for the rule Hypothesis-0074 is: 1.0
[360] Attribute SIMPLICITYof Hypothesis-0074-1 changed: (1.0 13 1.0)
[360] Discriminated rule: Hypothesis-0074-1
((GOAL: predict)
```

```
  (Exp =e Substance =iv1)
  (Exp =e Radius =iv2)
  (Exp =e ODist =iv3)
  (Exp =e Alpha =iv4)
  (Exp =e Theta1 =15.0))
[360] Attribute SIMPLICITYof Hypothesis-0074-2 changed: (1.0 13 1.0)
[360] Discriminated rule: Hypothesis-0074-2
((GOAL: predict)
  (Exp =e Substance =iv1)
  (Exp =e Radius =iv2)
  (Exp =e ODist -200.0)
  (Exp =e Alpha =iv4)
  (Exp =e Theta1 =iv6))
[360] Attribute SIMPLICITYof Hypothesis-0074-3 changed: (1.0 13 1.0)
[360] Discriminated rule: Hypothesis-0074-3
((GOAL: predict)
  (Exp =e Substance =iv1)
  (Exp =e Radius plane)
  (Exp =e ODist =iv3)
  (Exp =e Alpha =iv4)
  (Exp =e Theta1 =iv6))
[375] Attribute SIMPLICITYof Hypothesis-0094 changed: (1.0 12 1.0)
[375] Rule newly built:
Hypothesis-0094 (Hypothesis: DV: IDist IV: ODist Form: (LAMBDA (iv) (ROUNDED.QUOTIENT iv -.43)))
[375] Attribute SIMPLICITYof Hypothesis-0095 changed: (1.0 12 3.0)
[375] Rule newly built:
Hypothesis-0095 (Hypothesis: DV: IDist IV: ODist Form: (LAMBDA (IV) (ROUNDED.DIFFERENCE IV -333.0)))
[375] Attribute SIMPLICITYof Hypothesis-0096 changed: (1.0 12 1.0)
[375] Rule newly built:
Hypothesis-0096 (Hypothesis: DV: Gamma IV: Alpha Form: (LAMBDA (iv) (ROUNDED.QUOTIENT iv -2.22)))
[375] Attribute SIMPLICITYof Hypothesis-0097 changed: (1.0 12 3.0)
[375] Rule newly built:
Hypothesis-0097 (Hypothesis: DV: Gamma IV: Alpha Form: (LAMBDA (IV) (ROUNDED.DIFFERENCE IV 14.5)))
[375] Attribute SIMPLICITYof Hypothesis-0098 changed: (1.0 12 1.0)
[375] Rule newly built:
Hypothesis-0098 (Hypothesis: DV: Gamma IV: Theta1 Form: (LAMBDA (iv) (ROUNDED.QUOTIENT iv 4.56)))
[375] Attribute SIMPLICITYof Hypothesis-0099 changed: (1.0 12 3.0)
[375] Rule newly built:
Hypothesis-0099 (Hypothesis: DV: Gamma IV: Theta1 Form: (LAMBDA (IV) (ROUNDED.DIFFERENCE IV -16.0)))
[375] Attribute SIMPLICITYof Hypothesis-0100 changed: (1.0 12 1.0)
[375] Rule newly built:
Hypothesis-0100 (Hypothesis: DV: Theta2 IV: Alpha Form: (LAMBDA (iv) (ROUNDED.TIMES iv .57)))
[375] Attribute SIMPLICITYof Hypothesis-0101 changed: (1.0 12 2.0)
[375] Rule newly built:
Hypothesis-0101 (Hypothesis: DV: Theta2 IV: Alpha Form: (LAMBDA (IV) (ROUNDED.PLUS IV -4.3)))
[375] Rule: Hypothesis-0078 rebuilt.
[375] Attribute SIMPLICITYof Hypothesis-0103 changed: (1.0 12 2.0)
[375] Rule newly built:
Hypothesis-0103 (Hypothesis: DV: Theta2 IV: Theta1 Form: (LAMBDA (IV) (ROUNDED.PLUS IV 26.2)))
```

Experiment exp-22
```
[381]Design: diamond v:100.0 v:-100.0 v:15.0 v:-31.4
[385]Feedback: v:8.8 v:-7.0 v:233.0 W
[389] (Prediction-0104 ISA: Prediction DV: Theta2 IV: Theta1 Form: (LAMBDA (iv) (ROUNDED.TIMES iv -.28
)) Value: 8.79)based on hypothesis Hypothesis-0078 is correct
[389] The new attribute value for the rule Hypothesis-0078 is: 4.0
```

Experiment exp-23
```
[395]Design: diamond v:100.0 v: 200.0 v:10.0 v:-32.3
[399]Feedback: v:9.0 v:-12.64 v:175.0 W
[403] (Prediction-0105 ISA: Prediction DV: Theta2 IV: Theta1 Form: (LAMBDA (iv) (ROUNDED.TIMES iv -.28
)) Value: 9.04)based on hypothesis Hypothesis-0078 is correct
[403] The new attribute value for the rule Hypothesis-0078 is: 5.0
```

Experiment exp-24
```
[409]Design: diamond v:100.0 v:-200.0 v:10.0 v:-50.7
[413]Feedback: v:14.2 v:-20.52 v:175.0 W
[417] (Prediction-0106 ISA: Prediction DV: Theta2 IV: Theta1 Form: (LAMBDA (iv) (ROUNDED.TIMES iv -.28
)) Value: 14.2)based on hypothesis Hypothesis-0078 is correct
[417] The new attribute value for the rule Hypothesis-0078 is: 6.0
```

Experiment exp-25
[423] Design: glass plane v:-100.0 v:20.0 v:-20.0
[427] Feedback: v:8.0 v:8.0 v:-250.0 W
[431] (Prediction-0107 ISA: Prediction DV: Theta2 IV: Theta1 Form: (LAMBDA (iv) (ROUNDED.TIMES iv -.28
)) Value: 5.6)based on hypothesis Hypothesis-0078 is wrong
[431] The new attribute value for the rule Hypothesis-0078 is: 1.5
[431] Attribute SIMPLICITYof Hypothesis-0078-1 changed: (1.0 13 1.0)
[431] Discriminated rule: Hypothesis-0078-1
((GOAL: predict)
 (Exp =e Substance =iv1)
 (Exp =e Radius =iv2)
 (Exp =e ODist =iv3)
 (Exp =e Alpha 10.0)
 (Exp =e Theta1 =iv6))
[431] Attribute SIMPLICITYof Hypothesis-0078-2 changed: (1.0 13 1.0)
[431] Discriminated rule: Hypothesis-0078-2
((GOAL: predict)
 (Exp =e Substance =iv1)
 (Exp =e Radius =iv2)
 (Exp =e ODist -200.0)
 (Exp =e Alpha =iv4)
 (Exp =e Theta1 =iv6))
[431] Attribute SIMPLICITYof Hypothesis-0078-3 changed: (1.0 13 1.0)
[431] Discriminated rule: Hypothesis-0078-3
((GOAL: predict)
 (Exp =e Substance =iv1)
 (Exp =e Radius 100.0)
 (Exp =e ODist =iv3)
 (Exp =e Alpha =iv4)
 (Exp =e Theta1 =iv6))
[431] Attribute SIMPLICITYof Hypothesis-0078-4 changed: (1.0 13 1.0)
[431] Discriminated rule: Hypothesis-0078-4
((GOAL: predict)
 (Exp =e Substance diamond)
 (Exp =e Radius =iv2)
 (Exp =e ODist =iv3)
 (Exp =e Alpha =iv4)
 (Exp =e Theta1 =iv6))
[446] Attribute SIMPLICITYof Hypothesis-0108 changed: (1.0 12 3.0)
[446] Rule newly built:
Hypothesis-0108 (Hypothesis: DV: Gamma IV: Theta1 Form: (LAMBDA (iv) (ROUNDED.QUOTIENT iv -2.5)))
[446] Attribute SIMPLICITYof Hypothesis-0109 changed: (1.0 12 3.0)
[446] Rule newly built:
Hypothesis-0109 (Hypothesis: DV: Gamma IV: Theta1 Form: (LAMBDA (IV) (ROUNDED.DIFFERENCE IV -28.0)))
[446] Rule: Hypothesis-0031 rebuilt.
[446] Attribute SIMPLICITYof Hypothesis-0111 changed: (1.0 12 3.0)
[446] Rule newly built:
Hypothesis-0111 (Hypothesis: DV: Gamma IV: Alpha Form: (LAMBDA (IV) (ROUNDED.DIFFERENCE IV 12.0)))
[446] Rule: Hypothesis-0037-4 rebuilt.
[446] Attribute SIMPLICITYof Hypothesis-0113 changed: (1.0 12 3.0)
[446] Rule newly built:
Hypothesis-0113 (Hypothesis: DV: Theta2 IV: Theta1 Form: (LAMBDA (IV) (ROUNDED.PLUS IV 28.0)))
[446] Attribute SIMPLICITYof Hypothesis-0114 changed: (1.0 12 2.0)
[446] Rule newly built:
Hypothesis-0114 (Hypothesis: DV: Theta2 IV: Alpha Form: (LAMBDA (iv) (ROUNDED.TIMES iv .4)))
[446] Attribute SIMPLICITYof Hypothesis-0115 changed: (1.0 12 3.0)
[446] Rule newly built:
Hypothesis-0115 (Hypothesis: DV: Theta2 IV: Alpha Form: (LAMBDA (IV) (ROUNDED.PLUS IV -12.0)))
[446] Attribute SIMPLICITYof Hypothesis-0116 changed: (1.0 12 3.0)
[446] Rule newly built:
Hypothesis-0116 (Hypothesis: DV: IDist IV: ODist Form: (LAMBDA (iv) (ROUNDED.TIMES iv 2.5)))
[446] Attribute SIMPLICITYof Hypothesis-0117 changed: (1.0 12 3.0)
[446] Rule newly built:
Hypothesis-0117 (Hypothesis: DV: IDist IV: ODist Form: (LAMBDA (IV) (ROUNDED.PLUS IV -150.0)))

Experiment exp-26
[452] Design: flint v:100.0 v:-100.0 v:15.0 v:-31.4
[456] Feedback: v:10.4 v:-5.4 v:300.0 W
[460] (Prediction-0118 ISA: Prediction DV: Theta2 IV: Theta1 Form: (LAMBDA (iv) (ROUNDED.TIMES iv -.33
)) Value: 10.36)based on hypothesis Hypothesis-0052-1 is correct
[460] The new attribute value for the rule Hypothesis-0052-1 is: 3.0

Experiment exp-27
[466]Design: diamond plane v:-100.0 v:15.0 v:-15.0
[470]Feedback: v:4.2 v:4.2 v:-350.0 W
[474] (Prediction-0119 ISA: Prediction DV: Theta2 IV: Theta1 Form: (LAMBDA (iv) (ROUNDED.TIMES iv
-.33
)) Value: 4.95)based on hypothesis Hypothesis-0052-1 is wrong
[474] The new attribute value for the rule Hypothesis-0052-1 is: .75
[474] The new attribute value for the rule Hypothesis-0052-1 is: 1.75
[474] Attribute SIMPLICITYof Hypothesis-0052-1-1 changed: (1.0 14 1.0)
[474]   Discriminated rule: Hypothesis-0052-1-1
((GOAL: predict)
 (Exp =e Substance =iv1)
 (Exp =e Radius =iv2)
 (Exp =e ODist -100.0)
 (Exp =e Alpha 15.0)
 (Exp =e Theta1 =iv6))
[474] Attribute SIMPLICITYof Hypothesis-0052-1-2 changed: (1.0 14 1.0)
[474]   Discriminated rule: Hypothesis-0052-1-2
((GOAL: predict)
 (Exp =e Substance =iv1)
 (Exp =e Radius 100.0)
 (Exp =e ODist =iv3)
 (Exp =e Alpha 15.0)
 (Exp =e Theta1 =iv6))
[474] Attribute SIMPLICITYof Hypothesis-0052-1-3 changed: (1.0 14 1.0)
[474]   Discriminated rule: Hypothesis-0052-1-3
((GOAL: predict)
 (Exp =e Substance flint)
 (Exp =e Radius =iv2)
 (Exp =e ODist =iv3)
 (Exp =e Alpha 15.0)
 (Exp =e Theta1 =iv6))

APPENDIX VI.4

PRISM Code for Model HDD-QUAL

For information on the organization of the listing and the syntax of PRISM rules see Appendix VI.1.

{FSO:COGNITIVE SYSTEMS:PSYCHOLINSTITUT}<REIMANN>DISSGRAPHS>APDX-HDD-QUAL.;1

(FILECREATED " 9-Mar-89 15:46:09" {DSK}<LISPFILES>LISPUSERS>REIMANN>DISSGRAPHS>HDD-QUA
L.;1 17373

    changes to: (VARS HouseHold.PM GenDesign.2 ModifyH.2)

    previous date: " 5-Mar-89 16:20:02" {FLOPPY}<HDD>CODE>HDD-QUAL.;1)

(PRETTYCOMPRINT HDD-QUALCOMS)

(RPAQQ HDD-QUALCOMS ((ADDVARS (*MODULES* ENV-0.ARC ENVSpread.Arc WM-0.ARC CONTROL-0.ARC HM2.ARC
                              PHEN-1.ARC DISCRIM-0.ARC GEN-0.ARC HouseHold.PM
                              GenDesign.2 Predict&Compare.2 ModifyH.2 GenH.2 WM-1.CNT
                              HM1.CNT))
         (VARS ENV-0.ARC ENVSpread.Arc WM-0.ARC CONTROL-0.ARC HM2.ARC PHEN-1.ARC DISCRIM-0.ARC
               GEN-0.ARC HouseHold.PM GenDesign.2 Predict&Compare.2 ModifyH.2 GenH.2 WM-1.CNT HM1.CNT))
         (PROP FRAME HDD-QUAL)
         (P (SETQ *MODULES* (SORT *MODULES*))
            (PM.DataMenuHandler *MODULES*))))

(ADDTOVAR *MODULES* ENV-0.ARC ENVSpread.Arc WM-0.ARC CONTROL-0.ARC HM2.ARC PHEN-1.ARC DISCRIM-0.ARC
                    GEN-0.ARC HouseHold.PM GenDesign.2 Predict&Compare.2 ModifyH.2 GenH.2
                    WM-1.CNT HM1.CNT)

(RPAQQ ENV-0.ARC (PROGN (* * set up an external memory ENV)
                        (CREATE-COMPONENT ENV INSTANCE-OF DECLARATIVE-MEMORY ATTRIBUTES (ACTIVATION)
                                          DEFAULT-VALUES
                                          (1.0)
                                          REVISED-VALUES
                                          (NEW)
                                          SYNTAX
                                          (Exp))
                        (printout T T "Created Architecture for ENV" T)))

(RPAQQ ENVSpread.Arc (PROGN (* * Define a retrieval process for ENV)
                            (* The verbose version)
                            (* CREATE-COMPONENT SPREAD-TO-LIMIT INSTANCE-OF PROPAGATION-PROCESS
                                PROCESSES ((MODIFY-AND-ADD-TO WM ELEMENT))
                                DEFAULT-AMOUNT 1.0 SPREAD-FROM-ELEMENT ((NEW-AMOUNT (FTIMES AMOUNT
                                                                                           .8))
                                                 (PRINT-FROM ELEMENT AMOUNT)
                                                 (NEW-AMOUNT (FQUOTIENT AMOUNT (NUMBER-OF-SYMBOLS)))
                                                 (BLOCK-ELEMENT)
                                                 (SPREAD-TO-SYMBOLS)
                                                 (UNBLOCK-ELEMENT))
                                SPREAD-TO-SYMBOLS
                                ((NEW-AMOUNT (FTIMES AMOUNT (NUMBER-OF-TIMES)))
                                 (SPREAD-FROM-SYMBOL))
                                SPREAD-FROM-SYMBOL
                                ((PRINT-THROUGH SYMBOL AMOUNT)
                                 (NEW-AMOUNT (FQUOTIENT AMOUNT (SUM-OF ENV-ACTIVATION)))
                                 (BLOCK-SYMBOL)
                                 (SPREAD-TO-ELEMENTS)
                                 (UNBLOCK-SYMBOL))
                                SPREAD-TO-ELEMENTS
                                ((NEW-AMOUNT (FTIMES AMOUNT (ENV-ACTIVATION ELEMENT)))
                                 (IS-AMOUNT? (GREATERP AMOUNT .05))
                                 (NEW-AMOUNT (MIN AMOUNT 5.0))
                                 (STORE-AMOUNT AMOUNT)
                                 (PRINT-STORE ELEMENT AMOUNT)
                                 (SPREAD-FROM-ELEMENT)))
                            (* The nonverbose version)
                            (CREATE-COMPONENT SPREAD-TO-LIMIT INSTANCE-OF PROPAGATION-PROCESS
                                              PROCESSES ((MODIFY-AND-ADD-TO WM ELEMENT))
                                              DEFAULT-AMOUNT 1.0 SPREAD-FROM-ELEMENT
                                              ((NEW-AMOUNT (FTIMES AMOUNT .5))
                                               (NEW-AMOUNT (FQUOTIENT AMOUNT (NUMBER-OF-SYMBOLS)))
                                               (BLOCK-ELEMENT)
                                               (SPREAD-TO-SYMBOLS)
                                               (UNBLOCK-ELEMENT))
                                              SPREAD-TO-SYMBOLS
                                              ((NEW-AMOUNT (FTIMES AMOUNT (NUMBER-OF-TIMES)))
                                               (SPREAD-FROM-SYMBOL))
                                              SPREAD-FROM-SYMBOL
                                              ((NEW-AMOUNT (FQUOTIENT AMOUNT (SUM-OF ENV-ACTIVATION))

{FS0:COGNITIVE SYSTEMS:PSYCHOLINSTITUT}<REIMANN>DISSGRAPHS>APDX-HDD-QUAL.;1

```
                                                                    ))
                                          (BLOCK-SYMBOL)
                                          (SPREAD-TO-ELEMENTS)
                                          (UNBLOCK-SYMBOL))
                                         SPREAD-TO-ELEMENTS
                                         ((NEW-AMOUNT (FTIMES AMOUNT (ENV-ACTIVATION ELEMENT)))

                                          (IS-AMOUNT? (GREATERP AMOUNT .05))
                                          (NEW-AMOUNT (MIN AMOUNT 5.0))
                                          (STORE-AMOUNT AMOUNT)
                                          (SPREAD-FROM-ELEMENT)))
                            (PRINTOUT T T "SPREAD-TO-LIMIT defined" T)))

(RPAQQ WM-0.ARC (PROGN (* * Create a recency for WM)
                     (MD WM ATTRIBUTES (RECENCY)
                         DEFAULT-VALUES
                         ((WM-COUNT+))
                         REVISED-VALUES
                         (NEW))
                     (printout T "Declarative Memory WM redefined" T)))

(RPAQQ CONTROL-0.ARC (PROGN (* * Set up CONTROL Prefer more recent entries and more specific
                                   entries)
                          (* HDDENV.WmCycleStats 'WM)
                          (* HDDENV.WmCyclePlot)
                          (CREATE-COMPONENT CONTROL INSTANCE-OF PROCEDURAL-MEMORY MATCHES-AGAINST

                                          WM EVERY-CYCLE ((ORDER-BY-FIRST-ELEMENT WM-RECENCY)
                                          (SELECT-BEST GREATERP)
                                          (ORDER-BY-SUMMED-ELEMENTS WM-RECENCY)
                                          (SELECT-ONE-BEST GREATERP)
                                          (FIRE-PRODUCTIONS)
                                          (REFRACT-FIRED))
                                          EVERY-FIRING
                                          ((PRINT-FIRING)
                                          (PRINT-NAME)
                                          (MY-PRODUCTION-WATCHER)
                                          (EVALUATE)))
                                   (printout T T "Procedural Memory CONTROL defined" T)))

(RPAQQ HM2.ARC (PROGN (* * Hypotheses are ordered by simplicity, specifity and strength. Only one
                          is allowed to fire. Note that the rule Predict.JustOnce had to be added to

                          HM as a starting rule in order to assure a single H.)
                     (* HDDENV.HmCycleStats 'HM)
                     (* HDDENV.HmCyclePlot)
                     (CREATE-COMPONENT HM INSANCE-OF PROCEDURAL-MEMORY MATCHES-AGAINST WM
                                     ATTRIBUTES (STRENGTH SPECIFITY SIMPLICITY)
                                     DEFAULT-VALUES
                                     (1.0 (COUNT-CONSTANTS)
                                           1.0)
                                     REVISED-VALUES
                                     ((FPLUS OLD 1.0)
                                      OLD OLD)
                                     EVERY-CYCLE
                                     ((HDDENV.HmCycleStats 'HM)
                                      (HDDENV.HmCyclePlot)
                                      (ORDER-BY-PRODUCTION HM-STRENGTH)
                                      (SELECT-BEST GREATERP)
                                      (ORDER-BY-PRODUCTION HM-STRENGTH)
                                      (SELECT-RANDOMLY)
                                      (FIRE-PRODUCTIONS)
                                      (REFRACT-FIRED)
                                      (DELETE-FROM WM (GOAL: predict))
                                      (CALL CONTROL))
                                     EVERY-FIRING
                                     ((DIVIDER)
                                      (PRINT FIRING)
                                      (PRINT NAME)
                                      (EVALUATE))
                                     DEFAULT-ACTIONS NIL SYNTAX NIL GRAPHFLG T)
                     (printout T T "Procedural memory HM defined" T)
                     (* * Transfer process for HM)
                     (CREATE-COMPONENT HM-VANISH INSTANCE-OF TRANSFER-PROCESS TEST
                                     (LESSP (STRENGTH ELEMENT)
                                            .1)
                                     PROCESSES
```

{FSO:COGNITIVE SYSTEMS:PSYCHOLINSTITUT}<REIMANN>DISSGRAPHS>APDX-HDD-QUAL.;1

```
                        ((PRINT-TRANSFER)
                         (WRITECR UNBUILDING ELEMENT)
                         (UNBUILD-FROM HM ELEMENT)
                         (HDD.DribbleTransfer HM-VANISH ELEMENT)))
                  (* The current lower limit corresponds to two decreases starting from the
                     initial value .1)
                  (printout T T "Created Transfer Process HM-VANISH " T)))

(RPAQQ PHEN-1.ARC (PROGN (* * Architecture of PHEN, the memory that holds rules for phenomenon
                         induction. - Uses a threshold value to select consideration rules)
                         (CREATE-COMPONENT PHEN INSTANCE-OF PROCEDURAL-MEMORY MATCHES-AGAINST WM
                              ATTRIBUTES (STRENGTH)
                              DEFAULT-VALUES
                              (1.0)
                              REVISED-VALUES
                              (OLD)
                              EVERY-CYCLE
                              ((ORDER-BY-PRODUCTION PHEN-STRENGTH)
                               (ABOVE-THRESHOLD .9)
                               (FIRE-PRODUCTIONS)
                               (REFRACT-FIRED))
                              EVERY-FIRING
                              ((PRINT-FIRING)
                               (PRINT-NAME)
                               (MY-PRODUCTION-WATCHER)
                               (EVALUATE)))
                         (printout T T "Procedural Memory CONTROL defined" T)))

(RPAQQ DISCRIM-0.ARC (PROGN (CREATE-COMPONENT DISCRIMINATE INSTANCE-OF DISCRIMINATION-PROCESS
                              WORKING-MEMORY WM LEARNING-ACTIONS ((
                                POSITIVE-CONDITIONS))
                              SYMBOL-REPLACEMENT OFF LIMIT 1 SIGNIFICANT-SYMBOLS
                              (Exp GOAL: Substance Radius Alpha Theta1 Gamma Theta2

                              ODist predict)
                              SIGNIFICANT-SYMBOLS-TESTS
                              (EXP←LABEL←P)
                              PRESELECTION T SELECTION-SYMBOLS (Exp))
                         (printout T T "Created discrimination process: DISCRIMINATE " T)
                         (CREATE-COMPONENT ASSIGN-CREDIT INSTANCE-OF DESIGNATION-PROCESS MEMORY
                              WM PROCESS NIL)
                         (CREATE-COMPONENT INCREASE-ATTRIBUTE-VALUE INSTANCE-OF REVISION-PROCESS

                              PROCESS INCREASE ATTRIBUTE STRENGTH FUNCTOR PLUS
                              FACTOR 1.0)
                         (CREATE-COMPONENT DECREASE-ATTRIBUTE-VALUE INSTANCE-OF REVISION-PROCESS

                              PROCESS DECREASE ATTRIBUTE STRENGTH FUNCTOR TIMES
                              FACTOR .25)
                         (printout T T
 "Created designation processes: ASSIGN-CREDIT  INCREASE-ATTRIBUTE-VALUE  DECREASE-ATTRIBUTE-VALUE "
                              T)))

(RPAQQ GEN-0.ARC (PROGN (CREATE-COMPONENT GENERALIZE INSTANCE-OF GENERALIZATION-PROCESS
                              PROCEDURAL-MEMORY HM GENERALIZATION-STRATEGY VARIABILIZE
                              ACTION-STRATEGY ORDERED SIGNIFICANT-SYMBOLS
                              (Exp GOAL: Substance Radius Alpha Theta1 Gamma Theta2
                                Beta ODist predict)
                              SIGNIFICANT-SYMBOLS-TESTS
                              (EXP←LABEL←P))
                        (printout T T "Created generalization process: GENERALIZE " T)))

(RPAQQ HouseHold.PM [BUILD-IN CONTROL (* * HOUSEHOLD RULES * *)
                        [HH.Focus ((GOAL: focus on experiment =newfocus)
                              (ExpsInFocus: =oldfocus)
                              (ExpsInNoteBook: !nb)
                              (*NOT (*EQUAL =newfocus =oldfocus)))
                              -->
                              (($DELETE-FROM WM #2)
                              ($BIND =moveout (&LDIFFERENCE =oldfocus =newfocus))
                              ($BIND =movein (&LDIFFERENCE =newfocus =oldfocus))
                              ($WRITECR "** Removing " =moveout from Focus)
                              ($HDD.RemoveExpFromWM* =moveout)
                              ($WRITECR "** Moving " =movein into Focus)
                              ($HDD.AddExpToWM* =movein =nb)
                              ($ADD-TO WM (ExpsInFocus: =newfocus]
                        (HH.FocusDone ((GOAL: focus on experiment !exps))
                              -->
```

{FS0:COGNITIVE SYSTEMS:PSYCHOLINSTITUT}<REIMANN>DISSGRAPHS>APDX-HDD-QUAL.;1

```
                                    (($DELETE-GOAL WM #1])

(RPAQQ GenDesign.2 [BUILD-IN CONTROL [GenDesign2.SetSubGoals ((GOAL: run experiment))
                                                            -->
                                                            (($CALL CONTROL)
                                                             ($DELETE-GOAL WM #1)
                                                             ($PRISM.ChangeAttribute
                                                                       Predict.PredictionDone HM
                                                                                       STRENGTH .6)
                                                             ($ADD-GOAL WM (GOAL: design
                                                                                    experiment]
                                  (GenDesign0.Stop ((GOAL: design experiment)
                                                   (CurrentExperiment =e1)
                                                   (Last experiment =e1))
                                                   -->
                                                   (($DELETE-GOAL WM #1)
                                                    ($WRITECR No more experiments ... I stop)
                                                    ($HALT)))
                                  [GenDesign0.DesignExp ((GOAL: design experiment)
                                                        (CurrentExperiment =e))
                                                        -->
                                                        (($DELETE-GOAL WM #1)
                                                         ($DELETE-FROM WM #2)
                                                         ($BIND =next (&HDD.NextExperiment =e))
                                                         ($ADD-TO WM (CurrentExperiment =next))
                                                         ($ADD-GOAL WM (GOAL: get design values)
                                                                 (GOAL: focus on experiment (=next]
                                  (GenDesign0.GetValues ((GOAL: get design values)
                                                        (CurrentExperiment =e))
                                                        -->
                                                        (($HDD.ReadDesignValues.1 =e Design)
                                                         ($DELETE-GOAL WM #1)
                                                         ($ADD-GOAL WM (GOAL: predict])

(RPAQQ Predict&Compare.2 [BUILD-IN CONTROL [Pred&Comp0.MakePrediction ((GOAL: predict))
                                                                     -->
                                                                     (($CALL HM)
                                                                      ($ADD-GOAL WM
                                                                               (GOAL: evaluate
                                                                                       prediction]
                                         [Pred&Comp0.EvalPred ((GOAL: evaluate prediction))
                                                              -->
                                                              (($DELETE-GOAL WM #1)
                                                               ($ADD-GOAL WM (GOAL: compare prediction
                                                                                       feedback)
                                                                       (GOAL: get feedback]
                                         (Pred&Comp0.GetFeedbackValues ((GOAL: get feedback)
                                                                       (ExpsInNoteBook: !first =some)
                                                                       (CurrentExperiment =e)
                                                                       (*NOT (*EQUAL =some =e)))
                                                                       -->
                                                                       (($DELETE-FROM WM #2)
                                                                        ($ADD-TO WM
                                                                                (ExpsInNoteBook: !first
                                                                                        =some =e))

                                                                        (($HDD.ReadDesignValues.1 =e
                                                                                        Feedback)
                                                                         =e)
                                                                         ($DELETE-GOAL WM #1)))
                                         (Pred&Comp0.GetFeedbackValues1 ((GOAL: get feedback)
                                                                        (CurrentExperiment =e)
                                                                        (ExpsInNoteBook:))
                                                                        -->
                                                                        (($DELETE-FROM WM #3)
                                                                         ($ADD-TO WM (ExpsInNoteBook:
                                                                                         =e))
                                                                         ($HDD.ReadDesignValues.1 =e
                                                                                         Feedback)
                                                                         ($DELETE-GOAL WM #1)))
                                         [Pred&Comp0.Compare1 ((GOAL: compare prediction feedback)
                                                              (=p ISA: Prediction !form)
                                                              (CurrentExperiment =e))
                                                              -->
                                                              (($ADD-GOAL WM (GOAL: compare #2 =e]
                                         [Pred&Comp1.CorrectPred ((GOAL: compare
                                                                 (=p ISA: Prediction FORM: =form)
                                                                 =e)
```

```
{FSO:COGNITIVE SYSTEMS:PSYCHOINSTITUT}<REIMANN>DISSGRAPHS>APDX-HDD-QUAL.;1

                                        (Exp =e Ray =form)
                                        (FailureCount =n))
                                        -->
                                        (($WRITECR ***Prediction is correct)
                                         ($DELETE-GOAL WM #1)
                                         ($DELETE-FROM WM (FailureCount =n))
                                         ($ADD-TO WM
                                                  (Correct (=p ISA: Prediction
                                                                FORM: =form))
                                                  (FailureCount 0]
                          (Pred&Comp1.WrongPred ((GOAL: compare
                                          (=p ISA: Prediction FORM: =form)
                                          =e)
                                        (<NOT> (Exp =e Ray =form))
                                        (Exp =e Ray =form2)
                                        (FailureCount =n))
                                        -->
                                        (($DELETE-GOAL WM #1)
                                         ($DELETE-FROM WM (FailureCount =n))
                                         ($ADD-TO WM (DifferencePredFeedb:
                                                                  Predicted:
                                                                      =form
                                                                    FbValue:
                                                                     =form2)
                                                     (FailureCount (&ADD1 =n)))
                                         ($WRITECR ***Prediction is wrong)))
                          [Pred&Comp0.CompareDone ((GOAL: compare prediction feedback)
                                        (=p ISA: Prediction FORM: =form))
                                        -->
                                        (($DELETE-GOAL WM #1)
                                         ($ADD-GOAL WM (GOAL: eval hypotheses]
                          (Pred&Comp0.NoH ((GOAL: compare prediction feedback)
                                        (<NOT> (=p ISA: Prediction FORM: =form)))
                                        -->
                                        (($DELETE-GOAL WM #1)
                                         ($ADD-GOAL WM (GOAL: generate hypotheses]))
(RPAQQ ModifyH.2 [BUILD-IN CONTROL [ModifyH2.CorrectH ((GOAL: eval hypotheses)
                                                (Correct (=p ISA: Prediction FORM: =form)))
                                        -->
                                        (($WRITECR ***Hypothesis is correct, I
                                                        strengthen the rule)
                                         ($HDD.DribblePrediction (=p ISA: Prediction
                                                                    FORM: =form)
                                                                 (&GET-RULE-NAME
                                                                    =p)
                                                                 correct)
                                         ($STORE-CONDITIONS ($GET-CREDIT
                                                                (=p ISA: Prediction
                                                                     FORM: =form)))
                                         ($INCREASE-ATTRIBUTE-VALUE ($GET-RULE-NAME
                                                                        =p))
                                         ($DELETE-FROM WM #2)
                                         ($DELETE-FROM WM (=p ISA: Prediction FORM:
                                                                     =form]
                           [ModifyH2.WrongPred ((GOAL: eval hypotheses)
                                        (=p ISA: Prediction FORM: =form)
                                        (DifferencePredFeedb: Predicted: =form FbValue:
                                                                         =form2))
                                        -->
                                        (($WRITECR ***Prediction is wrong...I try to find
                                                        new conditions)
                                         ($HDD.DribblePrediction (=p ISA: Prediction FORM:
                                                                         =form)
                                                                 (&GET-RULE-NAME =p)
                                                                 wrong)
                                         ($DECREASE-ATTRIBUTE-VALUE (&GET-RULE-NAME =p))
                                         ($DISCRIMINATE (=p ISA: Prediction FORM: =form))
                                         ($HM-VANISH HM)
                                         ($DELETE-FROM WM #2 #3)
                                         ($ADD-TO WM (A hypothesis was wrong]
                           [ModifyH0.CallGenerator ((GOAL: eval hypotheses)
                                        (A hypothesis was wrong))
                                        -->
                                        (($DELETE-GOAL WM (GOAL: eval hypotheses))
                                         ($DELETE-FROM WM (A hypothesis was wrong))
                                         ($ADD-GOAL WM (GOAL: generate hypotheses]
                           (ModifyH0.NextExp ((GOAL: eval hypotheses)
```

{FS0:COGNITIVE SYSTEMS:PSYCHOLINSTITUT}<REIMANN>DISSGRAPHS>APDX-HDD-QUAL.;1

```
                                        (CurrentExperiment =e)
                                        (<NOT> (A hypothesis was wrong)))
                                        -->
                                       (($DELETE-GOAL (GOAL: eval hypotheses))
                                        ($ADD-GOAL WM (GOAL: run experiment])

(RPAQQ GenH.2 [BUILD-IN CONTROL (* * RULES THAT CREATE HYPOTHESES * *)
                        (GenH1.GenerateH ((GOAL: generate hypotheses)
                                         (CurrentExperiment =e)
                                         (Exp =e Ray =form))
                                         -->
                                        (($DELETE-GOAL WM #1)
                                         ($HDD.PredictionRule.1 =form HM)
                                         ($ADD-GOAL WM (GOAL: run experiment])

(RPAQQ WM-1.CNT (PROGN (* * Wm content for a qualitative learner using the horizontal line as
                              orientation * *)
                       (ADD-TO WM (Independent Substance)
                                  (Independent Radius)
                                  (Independent ODist)
                                  (Independent Alpha)
                                  (Dependent Ray)
                                  (Nominal Substance)
                                  (Nominal Radius)
                                  (Nominal ODist)
                                  (Nominal Alpha)
                                  (Nominal Ray))
                       (ADD-TO WM (Values Substance glass flint diamond)
                                  (Values Radius plane convex concave)
                                  (Values ODist close far)
                                  (Values Alpha up down)
                                  (Values Ray above-horizontal below-horizontal on-horizontal))
                       (ADD-TO WM (CurrentExperiment NIL)
                                  (ExpsInFocus: (exp-1))
                                  (ExpsInNoteBook:)
                                  (Last experiment exp-17)
                                  (FailureCount 0)
                                  (GOAL: run experiment))
                       (printout T T "Declarative memory WM initialized with contents of WM-1.CNT"
                               T)))

(RPAQQ HM1.CNT (BUILD-IN HM (* * HYPOTHESES MEMORY. It's here where the learned hypotheses rules
                                  will be stored and managed. * *)
                         (* * RUNEXP.PredictionDone::DESCRIPTION - IF the goal is make a prediction

                            and there is no hypothesis rule that can fire or all of them fired THEN

                            delete the prediction goal and return control to PM.)
                         (Predict.PredictionDone ((GOAL: predict))
                                                  -->
                                                 (($DELETE-GOAL WM #1)
                                                  ($CALL CONTROL)))
                         (* * Predict.JustOnce - is added as an additional stop rule to assure that

                            just one hypotheses will fire. If problem occur with this rule during
                            CS, its specifity value should be set to be very high (like 100))))

(PUTPROPS HDD-QUAL FRAME [HDD-QUAL (DOC (* * A qualitative learner))
                                   (A-KIND-OF (VALUE NO.BUCK.BRIGD.))
                                   [ENV-Arc (VALUE (ENV ARC (ENV-O.ARC ENVSpread.Arc]
                                   (WM-Arc (VALUE (WM ARC WM-O.ARC)))
                                   (CONTROL-Arc (VALUE (CONTROL ARC CONTROL-O.ARC)))
                                   (PHEN-Arc (VALUE (PHEN ARC PHEN-1.ARC)))
                                   (HM-Arc (VALUE (HM ARC HM2.ARC)
                                           (DOC (* changed to L2]
                                   (DISCRIMINATION (VALUE (DISCRIMINATE ARC DISCRIM-O.ARC)))
                                   (GENERALIZATION (VALUE (GENERALIZE ARC GEN-O.ARC)))
                                   (CONTROL-Content (VALUE (CONTROL PROC HouseHold.PM)))
                                   (GenDesign-Content (VALUE (CONTROL PROC GenDesign.2))
                                           (DOC (* Changed L2)))
                                   (Predict-Content (VALUE (CONTROL PROC Predict&Compare.2)))
                                   (ModifyH-Content (VALUE (CONTROL PROC ModifyH.2)))
                                   (GenH Content (VALUE (PHEN PROC GenH.2)))
                                   (ENV-Content (VALUE NIL))
                                   (WM-Content (VALUE (WM DECL WM-1.CNT)))
                                   (HM-Content (VALUE (HM PROC HM1.CNT))]
(SETQ *MODULES* (SORT *MODULES*))
(PM.DataMenuHandler *MODULES*)
```